TREES OF CALIFORNIA

Big-tree cone.

Trees of California:
Restored Enlarged Retro Special Edition

Written by Willis Linn Jepson
Originally printed in 1923

Oceanography and Sailing in the 19th Century: The Water World

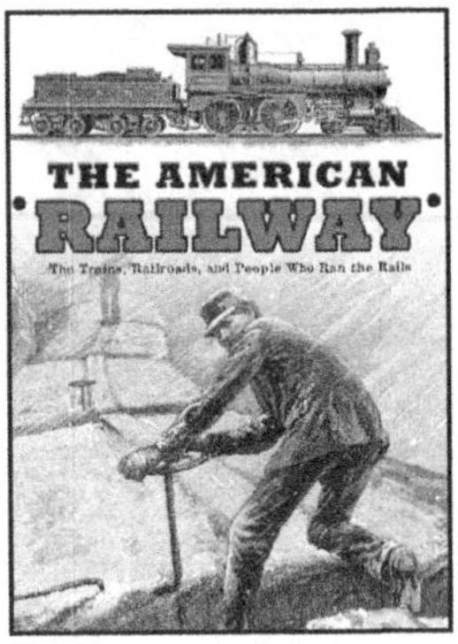

The American Railway: The Trains, Railroads, and People Who Ran the Rails

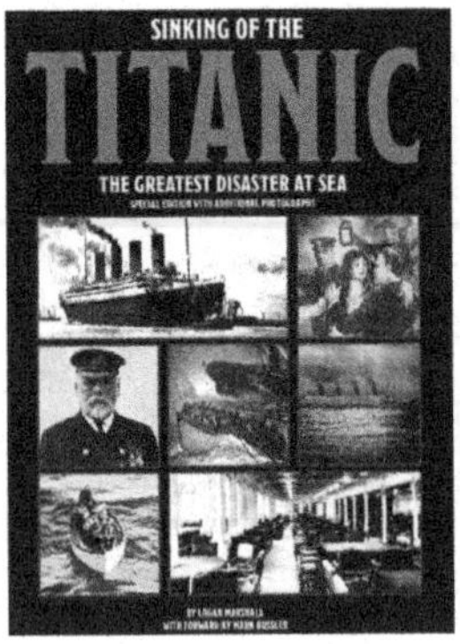

Sinking of the Titanic: The Greatest Disaster at Sea - Special Edition

THE TREES OF CALIFORNIA

BY

WILLIS LINN JEPSON

PROFESSOR OF BOTANY IN THE UNIVERSITY OF CALIFORNIA PRESIDENT OF THE CALIFORNIA BOTANICAL SOCIETY FELLOW OF THE ROYAL SOCIETY OF ARTS (LONDON) COUNCILLOR OF SAVE THE REDWOODS LEAGUE EDITOR OF ERYTHEA

ILLUSTRATED WITH ONE HUNDRED AND TWENTY-FOUR ORIGINAL DRAWINGS

SECOND EDITION

Fig. 1. BIG-CONE PINE (*Pinus coulteri* Don.). *a*, Cone. ½ nat. size. This species bears the largest and heaviest cones of any pine. The cones are well marked by the long talon-like appendages or curving spurs to the scales. See page 42. (Drawn by Miss Mary H. Swift.)

To My Friends

Frederick Folger Thomas, President of the Gwin Mine on the Mother Lode; William Anderson Scott Foster, President of the Northwestern Redwood Company; Ralph Hopping, naturalist on the North Fork of the Kaweah; Charles Russell Johnson, President of the Union Lumber Company; Carl Purdy, student of the Coast Range chaparral; Knowles Ryerson, wise guide to our youth of orchard and ranch; Ralph Platt, notary on the edge of the Vaca Mountains, but more than that, born naturalist—

To these, this book is dedicated in appreciation of their helpful aid and in memory of days and nights in mining camps in the cañon, logging camps in the forest, and pack-train camps on the mountain trails of Alta California.

"Monument" on a trail.

Pack saddle and Kyak.

PREFACE

Dr. James Bryce, British Ambassador to this country, once addressed informally a body of students at the University of California on the conduct of life. After speaking of those things necessary to real success in life he urged his hearers each to cultivate some interest beyond their life work or profession, like the study of some branch of botany, zoölogy or geology, which would serve as an intellectual recreation and as a resource from excessive cares of the day's or week's work.

The advice, while not new, was happily given. The lack of popular interest in the natural history sciences, failing some other cultivated interest, is unfortunate both for the individual and for the community. While this book from the standpoint of utility is designed primarily to provide a working manual of the native trees in small compass for use in the field by the horticulturist, farmer, cattleman, lumberman, mountaineer, forester, teacher or traveler who wishes to learn something of the botany of California trees, their names and their geographic and economic interest, it is also given out with still another purpose.

The natural surroundings of Californians are singularly rich and varied. A scientific interest in at least certain features of our natural environment, as for example the trees, shrubs or herbaceous plants, directs one to useful and agreeable intellectual activity. Accurate and detailed knowledge of even a small area lifts the possessor out of the commonplace and enables him directly or indirectly to contribute to the well-being and happiness of his community.

The author, therefore, cherishes the hope that these pages may be an inspiration to some who have opportunity to take up special studies of our trees for the sake of the intellectual pleasure and cultivation to be derived from such an avocation. The number and diversity of the native trees of California, their habits, places of growth, times of seeding, relation to different soils, reaction to fire and a host of such matters offer a most attractive field to the botanist. These things do not form a very "practical" study to be sure, but they are the basis of other things which are "practical"; and such study, moreover, offers a means of mental enjoyment which is cultivation in the best sense. In spite of our worship of the "practical" it is being more widely recognized that the cultivated man with keen intelligence and a broad and liberal outlook is getting more out of life and is really more practical after all than the so-called practical man who has narrowed his interests to those which concern his immediate personal needs, who is not stirred by the lure of the unknown, and who has "locked his door against the ideals" and imaginations of humanity.

This book, then, makes appeal to those who would enjoy the botany of the native trees and, perhaps challenged to explore the mysteries of their relationships, discover a world of interest in all those matters which serve to contribute to their classification—for their proper classification, a much desired object, is in reality an illuminating and organized compendium of their detailed structure, their congenital ties, their life history and their ancestry.

WILLIS LINN JEPSON.

UNIVERSITY OF CALIFORNIA,
Berkeley.
April 20, 1923.

CONTENTS

Fig. 3. A logged REDWOOD area. The standing trees are "culls," with second-growth below them.

THE TREES OF CALIFORNIA

Redwood

Figs. 3, 4, 5

Redwood (Sequoia sempervirens Endl.) inhabits mountain slopes, cañon sides and valleys near the sea and situated in the summer fog belt. While the main bodies of it are always found in proximity to the ocean, nevertheless it is not tolerant of ocean winds and prefers slopes or cañons or valley floors more or less protected or sheltered, and not exposed to the direct impact of ocean gales. For this reason it less commonly occurs on hillsides squarely facing the sea. The earliest navigators skirting the coast line, saw, as the ocean traveler sees today, barren hills, barren almost without interruption, rising from the beaches along the whole California shore. There is nothing to suggest that great forests grow behind those long stretches of hills which by their tan-colors and paucity or lack of arborescent vegetation betoken an unwooded land.

Geographically the Redwood is distributed from the Santa Lucia Mountains north to the extreme corner of southwestern Oregon, embracing an area 450 miles long and 1 to 40 miles wide. The main body occurs in a well-defined belt which begins in Del Norte County and extends southward through Humboldt and Mendocino counties to southern Sonoma County, with a transverse break in the belt in southern Humboldt.

South of Sonoma County the Redwood occurs only in detached and irregular bodies in Marin County, the Oakland Hills and the Santa Cruz and Santa Lucia mountains. It ranges east to the Hoods Peak Range, grows on the west side of Napa Valley and reappears in a very small body on the east side of Howell Mt. in the Napa Range where it descends almost to the floor of Pope Valley, thus occurring on the water-shed of a tributary of the Sacramento River. This locality is furtherest of any from the ocean. The entire body lies in California with the exception of two very small groves just over the California line in Curry County, Oregon.

In Mendocino County Redwood is associated with Douglas Fir, Tan Oak, Coast Hemlock and Lowland Fir. In Humboldt and Del Norte counties, where it attains its finest development, it occurs in almost pure stands, other species being very subordinate or absent.

Fig. 4. REDWOOD (*Sequoia sempervirens* Endl.). *a*, Cone-bearing branchlet with usual type of foliage; *b*, seed. nat. size.

Magnificent bodies of Redwood, as yet untouched by the axe or only partially exploited, occur on the main Eel River, South Fork Eel River, Van Duzen River, Mad River, Redwood Creek, lower Klamath River and Smith River. The trees in these splendid forests are mostly mature or past maturity, 6 to 16 feet in diameter, 100 to 200 feet in height or taller, and yield 125,000 to 150,000 feet board measure per acre. Limited areas have produced as high as 200,000 to 500,000 feet board measure per acre and yields of 1½ million feet to the acre are on record.

The age of mature Redwood is 500 to 1300 years. It has not in this matter been subject to so much imaginative controversy as has the Big Tree.

The wood is light, soft, straight-grained, free from resin, works easily, keeps its shape well and is subject to slight shrinkage or warping after initial seasoning. It is almost marvelously even grained, and the ease and fineness of its straight splitting excite the admiration of mechanics. The autumnal part of the annual layer in the wood is thicker than in Big Tree, which accounts for the greater strength and toughness of Redwood. It is very light but relatively to weight it is one of the strongest of coniferous woods. Other remarkable characteristics are that it ignites tardily, chiefly because wholly free from resin, and that on account of its spongy character it absorbs water from a fire hose with great rapidity. California cities in which the buildings are largely constructed of Redwood are subject to a fire-control far superior to municipalities where resinous pine or fir prevail as building material. The old San Francisco, for instance, although in the main a city of frame Redwood houses, never had an extensively destructive fire until 1906 when the water supply was completely shut off by earthquake disturbance.

Redwood is logged and milled into lumber on a large scale. The Redwood cut amounted to 411,689,000 feet board measure in 1905 and in 1921 it amounted to about 500,000,000 feet board measure. The greater portion of the product is consumed in California. By means of it have been built the farms, factories, railroads, cities and towns of California in great degree. The species occurs in great forests, the belt is readily accessible by water transport, and by means of modern logging machinery and mills an abundant supply has always been available. Redwood was one of the first and has always been one of the most important assets in the building of the present-day civilization in our California. In one sense it outranked all other natural resources in its immediate as well as lasting importance. California might have done without her gold mines but not without the resources of the Redwood belt. It is not too much to say that, in consequence of all this, the importance of the Redwood lumber industry to the life and happiness of the human population and to the industrial welfare of this State has not been overestimated. From the very beginning of the American occupation the wood of Redwood has been intimately associated with the daily habits and experiences of the people.

The writer of these lines is a Californian. He was rocked by a pioneer mother in a cradle made of Redwood. The house in which he lived was largely made of Redwood. His clothing, the books of his juvenile library, the saddle for his riding pony were brought in railway cars chiefly made of Redwood, running on rails laid on Redwood ties, their course controlled by wires strung on Redwood poles. He went to school in a Redwood schoolhouse, sat at a desk made of Redwood and wore shoes the leather of which was tanned in Redwood vats. Everywhere he touched Redwood. Boxes, bins, bats, barns, bridges, bungalows were made of Redwood. Posts, porches, piles, pails, pencils, pillars, paving-blocks, pipe lines, sometimes even policemen, were made of Redwood. It is employed for the manufacture of shingles, for siding, for interior finish, for fences and for furniture. Its lustrous beauty lends itself to the chasteness of interior decoration, its durability to the plain mudsills of houses, sheds, and factories.

One of the most emphatic tributes to the economic value of Redwood is that new uses are constantly being discovered for it. We ship our choicest grapes to distant lands packed in Redwood sawdust. We replace steel water-conduits with Redwood. We supply Redwood doors to the Central American market because the white ant does not eat Redwood.

The economic uses of the wood are indeed almost numberless; it is used for a truly vast variety of purposes. Redwood enters directly or indirectly into almost every industry and activity in the State and affects practically every individual. From the day of coming into this world, from the cradle to the grave, Californians are in some way in touch with Redwood; and since it works with

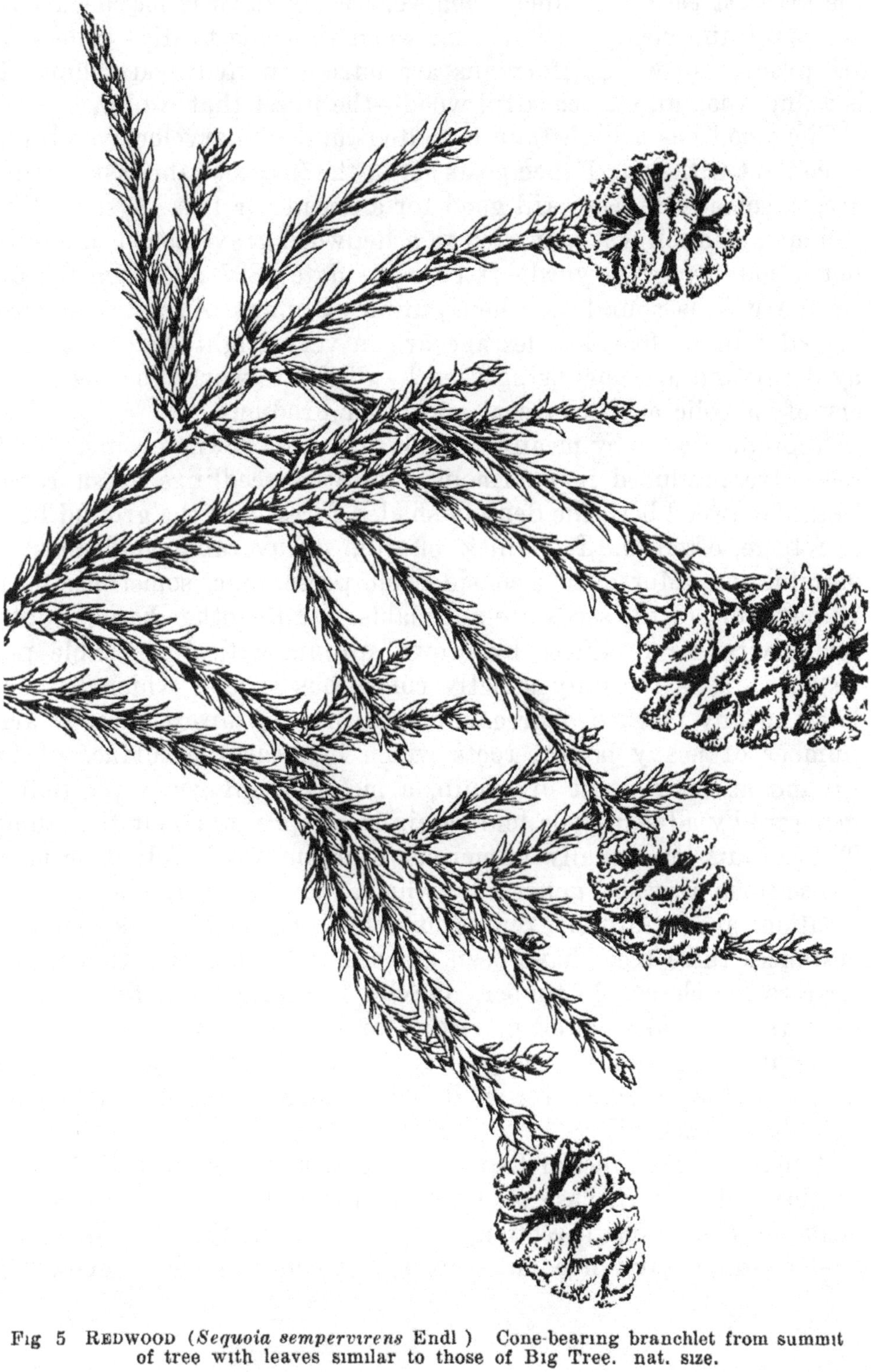

Fig 5 Redwood (*Sequoia sempervirens* Endl) Cone-bearing branchlet from summit of tree with leaves similar to those of Big Tree. nat. size.

the greatest ease and since from very early days it has been both abundant and cheap, so it is that when we come to die—unless we are profiteers—we Californians are buried in Redwood coffins. It is a fine wood and a beautiful wood—the finest that we have.

The wood has a high tannin content and is marvelously durable in contact with soil. Fence posts set in the fifties of the last century are, in some instances, still good for a decade or two. A short time ago my attention was attracted to a Redwood grave-board in a quiet north country grave-yard. It bore the date 1867 and save for fire ravage was as sound as when put in. Stumps of Redwood trees logged four or five decades ago are in very slight degree affected by decay and are now being used by distillate works for the recovery of carbolic acid, charcoal and other products.

Reproduction is by means of seeds and root-crown sprouts. While seeds are produced in enormous quantities seedlings are a rarity in the Redwood belt, the densely shaded forest and the ground litter of foliage, often one foot thick, offering unfavorable conditions for germination. Moreover a considerable percentage, sometimes 60 to 75 per cent., of the seeds are not viable. On the other hand Sequoia sempervirens reproduces by crown sprouts with remarkable persistence. It is the only strictly coniferous species which has this habit in any marked degree. The tree has no tap-root but a large number of heavy lateral roots which lie near the surface of the ground at their point of origin, a most advantageous position to generate by adventitious buds a circle of sprouts about the stump. The sprouts are usually numerous, sometimes a hundred or more. These form a second generation which, reduced in number by competition, are eventually represented by a circle of trees. In some instances these trees have been cut and a third generation is now near to merchantable timber. Circles resulting from the death of aged trees are also found in virgin forests. In most cases the original stump has wholly disappeared as a result of repeated fires and we have the shallow tree-encircled hollows which the woodsmen call "goose nests."

Admirable examples of Redwood circles may be seen in Mill Valley at the southeastern base of Mt. Tamalpais. One of these circles is fifty-one feet in one direction, forty-five feet in the other direction, and contains forty-five large trees, not counting small ones. The girth of the trunks ranges from less than two to six feet, while the crowns rise fifty to seventy feet. In the centre once stood the parent; the stumps represent the second generation, while the living trees are of the third generation. In this perfectly shaded circle divine services were at one time held and it is estimated that two hundred and fifty people could be comfortably seated within this natural enclosure.

The great handicap of the Redwood forest is fire as always in the case of a forest in California. The tree is well protected by the non-inflammable character of its very thick and fibrous bark and by its non-resinous wood. This forest as a whole is relatively immune since it lies in the area of highest seasonal rainfall in the State and this area is furthermore co-extensive with the summer fog belt. Nevertheless under certain combinations of seasonal dryness and wind the Redwood forest suffers severely from fire. Trees are, however, not always wholly consumed and it is no uncommon spectacle to see blackened masts, or even tall chimney-like trees 50 to 150 feet high, transformed in a short time into slender columns of soft-green foliage, so great is the regenerative capacity of this species.

The United States Forest Service estimates the original Redwood timber area as 1,406,000 acres. In 1922 the uncut stand consisted of about 951,000 acres, carrying about 70 billion feet of lumber board measure.

On account of the great value of the wood, the rapid growth of the tree and its remarkable capacity for regeneration, the species has great commercial possibilities in the way of sustained yield under silvical management of the woods and scientific lumbering—as opposed to a policy of destructive lumbering. In a natural stand on Big River the Division of Forestry of the University of California measured an acre plot which carried 263 second-growth trees about 60 years old. These trees, which arose from crown sprouts, averaged 142 feet high and scaled 138,000 board feet of lumber, thus showing the fastest volume growth of any soft-wood timber known. There is thus special inducement to care for the Redwood forest because of its power in crown sprouting and its unequaled volume growth.

Sir Dietrich Brandis, member (in charge of forests) of the Imperial Council for India, was at one time adviser of President

Roosevelt in framing a forest policy for the United States. He bore a curious hollow scar almost in the center of the forehead—a relic of a sword-thrust—which gave him a peculiar, indeed almost strange appearance. He was the original of the chief forester in Kipling's jungle tales—the tale of Mowgli[1]. My own happy days in London with this bronzed great-minded forester with the thunderous voice, kindly heart and courtly hospitality are a precious heritage. The year of leave in England drawing to a close I bade him good-bye; whereupon he drew his tall gaunt figure up to its full height and said: "Take this message to the people of California. Let California preserve her Redwoods so that there will be continuous replacement by new growth following continuous cutting. If only this be done she will be dowered with might and power. Under scientific forest management such an endowment to such a commonwealth will mean wealth and prosperity in the time to come."

The amount of standing timber in the Redwood belt has always been vast, measured by the needs of the present day. During many decades logging operations have been carried on without reference to a future crop of timber on the land. In the work of "falling" Redwoods, giant trees (after being laid out on the ground) were frequently rejected, in the old days of logging, because the grain of the wood did not meet the special requirements of the wood's foreman. After the trees are down and sawn into log lengths, fire was and still is almost universally used in the yarding process; that is, the area is cleared of brush and tops by fire, in order that the logging crews with stationary engines and wire cables may snake out the logs with the minimum of trouble. In other words under typical conditions an area is exploited and the land abandoned.

Very slowly during the last thirty years the idea of scientific forestry has been gaining foothold in California. A good many causes have been operating: the growth of public opinion, which is partly sentimental, but is also based on sound ideas of state economics; the development of the profession of forest engineering with the power to achieve results superior to those attained in the

1 See the story "In the Rukh" (Many Inventions, page 222), which is a sequel to the Second Jungle Book Sir Dietrich was very wonderful Read his wise words on forest policy and his delightful damnation of annual reports and such In the rukh he speaks to Mowgli, the jungle boy, now grown to be a man, under the name of Muller With scientific sagacity he apprehends the significance of Mowgli's actions in the jungle and brushes aside his forest warden's appeal to the mysterious

program of exploitation; the rapidly decreasing timber supply of the United States which is regarded as serious if not ominous. In this general change of view, the lumbering companies have been involved by the inexorable movement of the times, and brought to a gradual, if tardy change of policy.

There thus developed a crucial day in the life of the Redwood forest, and as may happen the opportunity and the man met. The situation was correctly sensed by David T. Mason, a trained and tried forester. Strong in sagacity, keen in conception, gifted with large vision and foresight he has at the decisive moment commanded the confidence and interest of the Redwood lumber companies, and today, as a result of his counsel, companies representing fifty-five per cent of the total Redwood product are definitely committed to a management of their woods with reference to perpetual yield. It is thus that the prophecy of Sir Dietrich Brandis is to be fulfilled by a young forester whose powers represent the combined elements of ordered imagination and technical skill.

Many unusual botanical characteristics culminate in Sequoia sempervirens; a sufficient number to make it one of the more remarkable trees of the earth's surface.—as it is in many respects the most remarkable tree of the cone-bearing class. It is the tallest tree on earth, reaching an extreme height of 340 feet, and is exceeded in bulk only by the Big Tree of the Sierra Nevada. It forms most prodigious stands of timber equaled by no other tree. It has to an exceptional degree immunity from disease, being largely free from wood-destroying fungi.[2] It has remarkable resistance to fire, both in the quality of its protective bark and in its wood. Its powers of vegetative reproduction are not remotely approached by any other conifer and its height growth is very rapid. Its longevity is imposing and attracts the greatest interest. Redwood, too, is remarkable for its racial continuity. It has probably changed little since Miocene times, and the colonies of today probably came down in their present form through the glacial periods and possibly through the Pleiocene epoch itself.

If a tree were to be placed on the Great Seal of California it would be the Redwood. Its longevity, its freedom from disease, its power to recover quickly from supreme disaster, its usefulness and

2 The decay pockets occasionally found in the trunks have not as yet been definitely connected with a specific fungus.

its beauty make it a symbol of the hopes and aspirations of the Californians. Probably few other living things in this State have obtained such complete mastery of an area as the Redwood in the localities of its best development. To our people it is the friendly Redwood and the most beloved tree in all the Californias.

Tens of thousands of people from the cities go to the Redwood groves of Marin, Sonoma and Mendocino for three or four months of the rainless season, setting up their household gods in the shelter of the second-growth circles and living freely in the open air amongst the cinnamon columns and under a green forest canopy. The beauty of the Redwoods is oft extolled. On the canvases of Keith and of Welch these trees have been immortalized; while many a poet has attempted to pay due praise in verse.

The inner side of the Redwood belt, away from the ocean, makes a singularly fine land. That is a forest of magnificent giants. Their slender columns, buff and red-brown, form true forest cathedrals with a canopy of foliage which is fern-like in its tracery. The eye wanders among the trunks, feeding contentedly on their lovely colors and perfect symmetry. In places one comes to trees standing a little apart as if to reveal the quality of their finished shafts and the charm of their bracketed foliage.

The botanical discovery of Redwood was made by Archibald Menzies, surgeon and botanist to the Vancouver Expedition. It is inevitable that Menzies saw the Redwood at one or more points on the California coast in 1792 or 1793, but 1794 is the date of his collection. We often wondered where he collected this specimen. One day the writer was examining the Menzies specimen in the Herbarium of the British Museum at South Kensington in London, when he chanced to turn over the sheet and there, written on the back, was the legend "Santa Cruz, Menzies." Although carried to England in 1796 the Menzies specimen remained undescribed until published as new in 1823 by Aylmer Bourke Lambert, an English botanist. Of all genera of conifers known to botanists at that time this Pacific Coast tree is most like the genus Taxodium or Bald Cypress. Lambert consequently placed the new species in that genus and, since it was evergreen, gave it the name Taxodium sempervirens to differentiate it from the previously known deciduous species, Taxodium distichum or Bald Cypress. The word sempervirens had no significance or connotation to Lambert other than as contrast to deciduous. Thus the species so remained until Stephen Endlicher, a German botanist, studying the classification of the conifers, decided that this species represented a genus distinct from Taxodium and created the new genus Sequoia in 1847, the Redwood thus becoming Sequoia sempervirens, a judgment which has received universal approval amongst botanists. The Redwood was therefore the first Sequoia to be made known, the character of the new genus having been published five years before the discovery of the second species, the Big Tree, Sequoia gigantea of the Sierras.

Big Tree

Fig. 6

Big Tree, the Sequoia gigantea of botanists and the Wawona of the Mokelumne Tribe, is a remarkable giant tree 125 to 331 feet high with columnar trunks 80 to 180 feet to the first limb and 5 to 25 feet in diameter at 6 feet above the ground. The crown in young trees is a regular pyramid resting on the ground; in the adult tree it is narrow with rounded summit; in old age it is more or less broken, typically with dead axis projecting above it.

Sequoia gigantea prefers slopes, ridges or depressions where there is sufficient moisture but it may grow on bare granite as in the Giant Forest. It is commonly associated with White Fir, Incense Cedar, Yellow Pine and Sugar Pine and never occurs in pure stands. It inhabits the western slope of the Sierra Nevada at 5000 to 8000 feet from Placer County southward to Tulare County, a longitudinal range of 250 miles, only occurring in more or less widely disconnected and limited areas called "groves" thirty-two in number. The northern groves, that is, north of Kings River, are widely separated; the southern groves, south of Kings River, are less widely separated or even connected by scattered individuals and form an interrupted belt[1].

[1] The northern groves are as follows: 1, North Grove, Placer Co. 2, Calaveras Grove. 3, Stanislaus Grove. 4, Tuolumne Grove. 5, Merced Grove. 6, Mariposa Grove. 7, Fresno Grove. 8. McKinley Grove, formerly known as Dinkey Grove. The southern groves are as follows: 9, Converse Basin Forest. 10, Boulder Creek Forest. 11, General Grant Forest. 12, Redwood Cañon Forest. 13, Muir Grove, formerly known as North Kaweah Forest. 14, Swanee River Grove. 15, Giant Forest. 16, Redwood Meadow Grove. 17, Harmon Meadow Grove. 18, Atwell Forest. 19, Lake Cañon Grove. 20, Mule Gulch Grove. 21, Homer Peak Forest. 22, South Kaweah Forest. 23, Dillon Forest. 24, Tule River Forest. 25, Pixley Grove. 26, Fleitz Forest. 27, Putnam Mill Forest. 28, Kessing Grove. 29, Indian Reservation Grove. 30, Deer Creek Grove. 31, Freeman Valley Forest. 32, Kern River Groves.

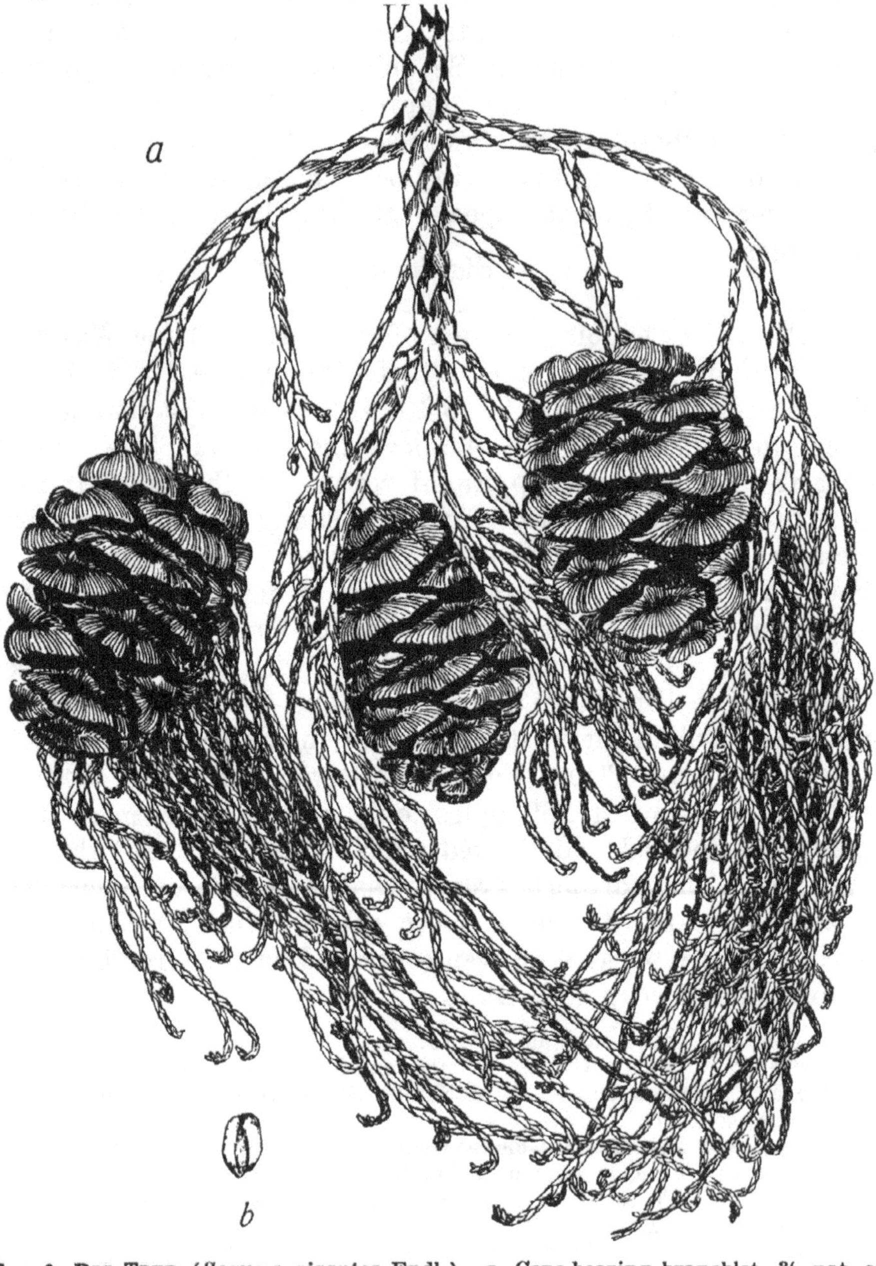

Fig. 6 Big Tree (*Sequoia gigantea* Endl). *a*, Cone-bearing branchlet, ⅔ nat size; *b*, seed, nat. size.

The Big Tree was discovered in the spring of 1852 by a hunter. one A. T. Dowd, whose pursuit of a grizzly bear led him into the Calaveras Grove. The discovery made some stir in the mining and lumber camps, was noised about in the local press, and news of it filtered down to San Francisco. Material came into the possession of the botanist, Dr. Albert Kellogg, before June, 1852, but he did not publish immediately concerning this remarkable species. The next year Kellogg showed his specimens of the Mammoth Tree, as it was then called, to William Lobb, the English botanical collector, recently arrived in California. Lobb immediately started for the Calaveras Grove, procured specimens and seedlings, and sailed from San Francisco on the first steamer for England. His material went to John Lindley, who created for this tree a new genus, which he called Wellingtonia, and a new species which he termed gigantea. Without delay he published it as Wellingtonia gigantea in December, 1853. The first to recognize the new conifer as a genuine species of Sequoia was J. Decaisne, the French botanist, who transferred it to that genus and published it as Sequoia gigantea in 1854. Kellogg and Behr, of San Francisco, somewhat patriotically incensed by Lindley's generic name, and unaware of Decaisne's paper, insisted on a reference of the species to Taxodium and published the Big Tree as Taxodium giganteum in 1855. The reference of the species to the genus Sequoia soon received, however, the critical approval of the best botanists of the time, an approval which has been fully established by all subsequent investigations of the comparative morphology of the Big Tree and Redwood.

The height of the Big Tree commonly averages from 125 to 225 feet, but trees in excess of these figures are well known. The best authenticated of recent figures of the extreme heights of known trees are as follows: The General Sherman Tree in the Sequoia Park is 279.9 feet high. The Dalton Tree in the Muir Grove is 292 feet high. These appear to be extreme figures in the forests of the southern Sierra Nevada. The Columbia Tree in the Mariposa Grove is 294 feet high and the Mark Twain Tree is said to be 331 feet high, a figure in excess of any measurements hitherto given which have been made by presumably accurate methods.

The dimensions in feet of a number of the more remarkable trees of the Mariposa Grove as given in the Department of the Interior Yosemite pamphlet for 1919 are shown in the following table:

DIMENSIONS OF SELECTED BIG TREES.

Name of Tree	Height	Diam. at 10 ft.	Diam. at base
Mark Twain	331	13.0	16.9
Capt. A. E. Wood	310	12.7	16.5
Columbia	294	16.5	25.6
South Carolina	264	17.3	23.5
Washington	235	20.7	29.3
Forest Queen	219	12.1	17.0
Grizzly Giant	204	20.5	29.6

Height is a matter which can be determined with a fair degree of accuracy and when determined is definite. Diameter is, however, in different case. Diameters at the ground do not, in many cases, give a significant value to the trunk, for the reason that the Big Trees often swell excessively at base, and the diameter at the base by actual measurement I have found in certain cases to be twice that at 10 feet above the ground. The only figures valid for purposes of comparisons must therefore be taken sufficiently above the ground to minimize error due to this factor. As so many persons have a natural interest in the largest known diameters there is here given the diameter of four of the most famous trees.

TABLE OF BIG TREE DIAMETERS.

Name of Tree	Diameter at 12 ft.	Diameter at ground
General Sherman (Sequoia Park)	27½ ft.	34⅓ ft.
General Grant (Grant Park)	23 ft.	35¾ ft.
Grizzly Giant (Mariposa Grove)	20 ft. (at 11 ft.)	31⅙ ft.
Boole Tree (Converse Basin)	25¾ ft. (at 10 ft.)	36 ft.

It must be emphasized, however, that characteristically the trunk taper is very slight. Indeed all observers unite in agreeing that the feature of a Big Tree, more remarkable and impressive than any other, is the columnar character of its trunk, the great height of the clear column and the manner in which it maintains its diameter upward from the ground to the crown.

The tree is furnished with a deeply furrowed bark, which is very non-inflammable, and which, by its peculiar fibrous nature and non-resinous quality, forms an almost asbestos-like covering to the trunk. This bark is of a beautiful red-brown or cinnamon color six to twelve inches thick. It often attains greater thickness, since bark two feet in thickness is actually known.

Fire burns through this heavy layer of bark very slowly and it is only after repeated forest fires that entrance is gained to the woody layers. Even then the rate of progress is not rapid because the wood is non-resinous and burns slowly. Nearly all mature trees or trees past maturity show some signs of fire or fire ravage, although in many cases the attack has been negligible.

The wood is dark red, but pink when freshly sawn, light and fairly strong. It is extraordinarily durable; posts last indefinitely and logs buried naturally, which must be centuries old, often show little or no decay in heartwood. Thousands of Big Trees on the Tule River, East Fork of the Kaweah River, at the Fresno Grove, and especially in Converse Basin have been logged and manufactured into lumber which is used for the same purpose as Redwood. The mature wood is without resin-ducts which are always absent from the wood except in the first annual ring of new growth in adult (cone-bearing) trees.

In appearance this wood does not differ very much from the well known Redwood of the coast. It has much the same color, texture and weight. The difference in strength can well be illustrated by observing in Tulare County vineyards grapevine stakes two inches square made from Big Tree logs and similar stakes about one inch square in Napa Valley vineyards made from Redwood. Redwood resists lateral strain, the Big Tree wood in a far less degree. Big Tree wood has a tendency to fracture transversely when split, whereas the Redwood splits cleanly lengthwise. One sometimes sees in the beds of Sierran rivers huge but short logs broken off sharply at the ends. They suggest great Leviathans, which have been rolled from side to side of the cañon walls and weathered by successive floods. They are fractured segments of the trunks of our great Sierran trees. No Redwood tree on the contrary would fracture in that way. Redwood trees split longitudinally.

As the tree goes on past maturity it eventually begins to die in the top, which may result from either of two causes: The gradual exhaustion of its food supply or years of deficient seasonal rainfall. It is possible and rather probable that the tops of Sequoias may be killed by lightning, but we are not able to give direct evidence to that effect. Certainly no Big Tree has ever been killed by lightning, although pines, firs and other trees of the Sierran forest have been so killed. Probably all old trees of Sequoia gigantea have been struck by lightning, certainly very very many have been within the period of the white man's observation of them. One of the most remarkable of forest experiences is to see at night a fire burning 150 or 200 feet in the air in the very top of a great Sequoia gigantea tree. Such fires are set by lightning. On account of their obvious inaccessibility and their tendency to throw off live sparks they are a source of worry to the forest ranger, who can do nothing but camp in the neighborhood until the fire burns itself out or until a propitious rainfall extinguishes it.

The cones of the Big Tree are very small (1½ to 2¾ inches long) —so small as to provoke comment on the contrast between size of tree and cone. They are produced in abundance and contain about 200 to 300 seeds, a sufficient number of which are viable. In the southern groves seedlings appear in large numbers, especially on "burns" or fire-spots still protected by the mother forest. Seedlings and young trees of every age may be seen in the Sequoia Park. In the northern groves reproduction is less common but some seedlings occur in most of the groves.

The age of Sequoia gigantea averages 900 to 2100 years so far as certainly known. The trees whose ages are certainly known are those which have been logged. When one considers that the average of the oldest logged trees were seedlings three hundred years before the Christian era it would seem that such a lengthened period of life were sufficient to afford ample food to the reflective mind. But those popular writers, and eke the poets; whose figures are based solely upon an admiring contemplation of the bulk and stateliness of these forest giants are not satisfied with attributing to them ages less than 5000 to 10,000 years.

Standing in rhapsodical admiration before a Sequoia gigantea one can easily imagine it to be 5000 to 10,000 years of age. The figure 8000 years has been placed on the Grizzly Giant, a famous tree in the Mariposa Grove, at the instance of California's most distinguished authority on fishes. As a matter of fact no one knows the age of the Grizzly Giant, as there is no satisfactory way of determining its age except by cutting it down. A small core could be taken from its trunk by a special tool, but this means might not prove satisfactory and such mutilation is not likely to be permitted.

There is no way of determining the age of a particular individual by means merely of the diameter. From various age studies of logged trees the writer finds, on the average, about 20 years to the inch.

Let us apply these figures to a particular individual. The Father of the Forest in the Calaveras Grove has a diameter of 27 feet inside the bark at about 8 feet above the ground. Its calculated age would therefore be 6840 years. When cut down its age was determined to be about 1300 years. In the Converse Basin I determined the age of a tree 11 feet 7½ inches in diameter to be 2019 years. That is to say my figure is accurate in the sense that there is a possible error of only 10 or 15 years in either direction. Another tree 24 feet in diameter, that is to say, twice the diameter of the tree last mentioned, was only 1346 years old, that is to say, a little over half the age. Trees of various species often take on an appearance of great age when comparatively young, due to extremes of temperature, wind, disease or under-nutrition. Senility or its appearance is not always a matter of years, and the attempts to give a longevity of 4000 years or more to the Big Tree rest as yet on no substantial basis.[3]

Sequoia gigantea is, however, sufficiently old. It is indeed the oldest living thing upon the planet. In not a few cases it probably attains to an age of 3000 years. The age of one tree, logged in the Converse Basin, has been determined with closely approximate accuracy as 3148 years. This is the oldest tree of which we have a definite age record.

In much of the popular literature upon Sequoia gigantea it is the habit to speak of this tree as passing out, as a relic, as making its last stand upon the western flanks of the Sierra Nevada, as being a decadent survival. In a sense it is a survival but it is a most vigorous survival. No other tree grows to so great a size, no other tree has such longevity, no other coniferous tree has such resistance to disease excepting only its near relative, the Redwood of the coast. In open spaces in the forest seedlings appear in great numbers, especially in the southern Sierra Nevada, often forming weedy thickets which are sometimes very dense. In the same region too it sometimes forms extensive forests and is often the dominant tree in its areas of best development. It is so abundant that it has been lumbered on an extensive scale and millions of board feet of

[3] "Certainly the only way of giving reliable accounts on so exceptional a tree as Sequoia gigantea is to accept only numbers well ascertained, and not those which may be influenced by even an involuntary exaggeration caused by enthusiasm not controlled by competence."—Dr. F. Mader, Nice, France, in a letter to the author, 28 Jan., 1914.

lumber from the Big Tree have been milled and marketed as Redwood.

The Big Trees are remarkable forest products. In stature they are imposing as no other living thing; in age they are a measure for the centuries; in situation they are restricted to one slope of a mountain range where they are not able to extend appreciably their area; in number they are comparatively few; and in genealogy they are the direct descendants of a family dominant in the Tertiary period and richer then in genera and species than now.

On account of the unequaled character of the pines and firs in the Sierran forest the first sight of the Big Trees may be disappointing. But association with the Sequoias affords unusual opportunity for the intellectual traveler. As the days in their company run fleetly by, his appreciation of their scientific interest continues to deepen and strengthen until he instinctively senses something of the part they and their extinct congeners played in the Miocene epoch of the Tertiary.

Sugar Pine

Fig. 7

Sugar Pine (Pinus lambertiana Dougl.) is a splendid forest tree 70 to 180 feet high. The trunk, which is 3 to 7 feet in diameter, holds its diameter well upward, is usually clear of branches for a great height, and in typical trees is surmounted by a flat-topped or irregular crown consisting of several horizontal arms of unequal length, characteristics which distinguish it from all associated species. The bark is brown or reddish, 2 to 4 inches thick, fissured longitudinally into rough ridges, the surface breaking down into small deciduous scales. The slender needles are borne 5 in a cluster, while its wonderful cones are the longest of any pine.

Pinus lambertiana attains its greatest development in the main timber belt of the Sierra Nevada where it is, on account of its charmingly irregular crowns, a striking feature of the forest between 3500 and 6500 feet at the north and 5500 and 8500 feet at the south. While usually forming but a small portion of the forest stand it is in limited areas the dominant species. The fine Sugar Pine forests about Crocker's station and Hazel Green have been greatly admired by travelers journeying over the old-time Big Oak Flat and Coulterville wagon roads to Yosemite.

In the Coast Ranges it is comparatively scarce, being found only in the highest ranges. It does, however, extend nearly to the coast in Sonoma County, where it is associated with Redwood. In Southern California it is found generally on all the high ranges from the San Emigdio Mountains to the San Gabriel, San Bernardino, San Jacinto and Cuyamaca mountains. It recurs on Mt. San Pedro Martir in Lower California (southermost locality). Northward it ranges to the Santiam River in southern Oregon.

Sugar Pine wood is soft, light, close- and straight-grained, very white and satiny when finished, and of high commercial value. This species does not, however, renew its stands naturally with the same facility as Yellow Pine, Incense Cedar, or White Fir, and on account of the great demand for its wood the tree in the forest is rapidly becoming scarce. Sometimes superior logs of other species are marketed designedly or inadvertently as Sugar Pine. My former student, Vance S. Brown, has supplied me the following note: "After the bark has been wet or discolored by the water in the pond, the top logs and the logs from young trees of Yellow Pine and Jeffrey Pine greatly resembled Sugar Pine logs. In fact, the most expert scalers are often mistaken as to the species of some logs. In the great majority of cases the ends of Sugar Pine logs will show a solid smear of pitch covering the sap wood. In the case of donkey logs the ends are sometimes broken and covered with dirt so that this point cannot be observed. If any bark has been left on the log, whether discolored or not, the lower layers of scales on a Sugar Pine log will have a narrow scale-like fringe bordering each layer. This fringe is really a part of the scale and is easily detached. I have never observed this in any Yellow Pine."

Sugar Pine derives its name from the sugary exudation, sought by the native tribes, which forms hard white crystalline nodules on the upper side of fire- or axe-wounds in the wood. This flow contains little resin, is manna-like, has cathartic properties and is as sweet as cane-sugar, but belonging to a different class of sugars, namely pinite or pine-sugar. The nodules are soluble in water, nearly insoluble in alcohol and not fermentable.

The Sugar Pine and the name of David Douglas, botanical explorer, are inseparably linked. From 1824 to his death in 1833 Douglas was in the service of the London Horticultural Society. A born enthusiast, hardy of body, self-reliant and highly resourceful

—especially in meeting hostile or tricky savages—he explored portions of Oregon and Washington and later of California. In his journal, a highly entertaining narrative, is told the story of his adventurous journey from the Columbia River to the headwaters of the Umpqua River in search of the Sugar Pine, the seeds of which he had seen in the tobacco pouch of an Indian on the Columbia. Douglas recognized the species as new and named it in honor of his friend, Aylmer Bourke Lambert, a founder of the Linnean Society of London and the author of a sumptuous work on pines.

Silver Pine

Fig 8

The Silver Pine (Pinus monticola Don), also called Western White Pine and Mountain Pine, is a forest tree 50 to 125 feet high with a trunk 1 to 4 feet in diameter. The branches are slender, somewhat drooping, or mainly horizontal, especially above, the very tip-top with a cluster of ascending or semi-upright cone-bearing branches. The whitish or reddish bark is thin, very smooth or checked into small square or rectangular plates. The foliage is blue-green and somewhat glaucous. The wood is light, soft, close- and straight-grained.

In California, Silver Pine occurs mainly in the Sierra Nevada where it is found between 5500 and 8000 feet at the north and 8000 to 10,000 feet at the south. While widely scattered through the upper portion of the main timber belt, it is forestrally a rare tree and nowhere abundant except in small patches. It also occurs sparingly in the far North Coast Ranges (Trinity, Marble and Siskiyou mountains). Northward it ranges to British Columbia and Montana. Closely related to the Sugar Pine, it is sometimes called Little Sugar Pine.

White-bark Pine

Fig 9

White-bark Pine (Pinus albicaulis Engelm.) is a tree growing at or near timber line. In exposed situations at its upper limits it often forms broad mats hugging closely to rocky ridges; or again it is dwarfish and no more than 6 to 10 feet high. Trees with prostrate trunks 10 feet long and 1 to 1½ feet in diameter, or with crowns on the ground like low flat tables 2 or 3 feet high over which one may readily walk, are a characteristic feature of exposed ridges or huge cirques where snow-drifts 100 to 500 feet in

Fig. 9. WHITE-BARK PINE (*Pinus albicaulis* Engelm.). *a*, Closed cone; *b*, seed. nat. size.

depth accumulate in the winter. In protected cañons it may become 40 feet high with a single main axis from the base and ½ to 2 feet in trunk diameter. The bark is thin, whitish, smooth, or on old trunks fissured into scaly plates.

As a sub-alpine tree it is a feature of the high Sierra Nevada between 7000 to 11,000 feet at the north and 9500 to 12,000 feet at the south. The southernmost localities are about the head of Little Kern River and on Mt. Whitney. It is frequent on Bubbs Creek, about Bullfrog Lake, Tuolumne Meadows region (Vogelsang Peak, Mts. Lyell, Dana and Gibbs, and Benson Pass), and so on north to Mt. Shasta and west to Thompson Peak in Trinity County. It ranges far north to British Columbia and east to the Rocky Mountains of Montana.

The stand of White-bark Pine is very thin on account of uncertain reproduction. Cones are sparingly produced. Furthermore it is but rarely that one finds a good mature cone full of seed, on account of the industry of the crows and sub-alpine squirrels.

Limber Pine

Fig. 10

Limber Pine (Pinus flexilis James) is a tree 20 to 60 feet high with a short thick trunk 1 to 3 feet in diameter. The bark is dark brown, deeply furrowed and broken crosswise into nearly square plates. The branches are usually very long and long-persistent, extending down to or nearly to the ground. The foliage is dark yellow-green.

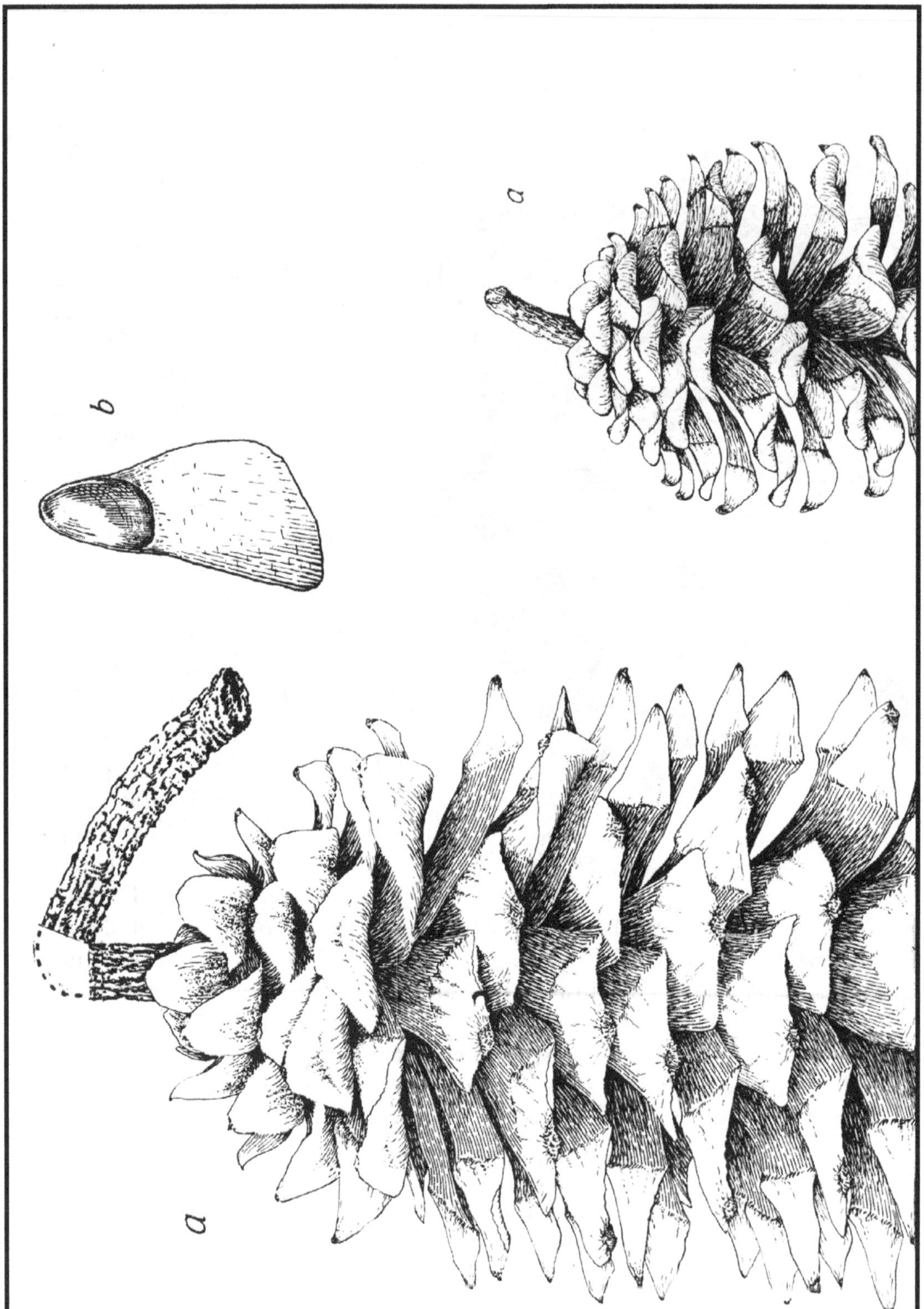
a
b
a

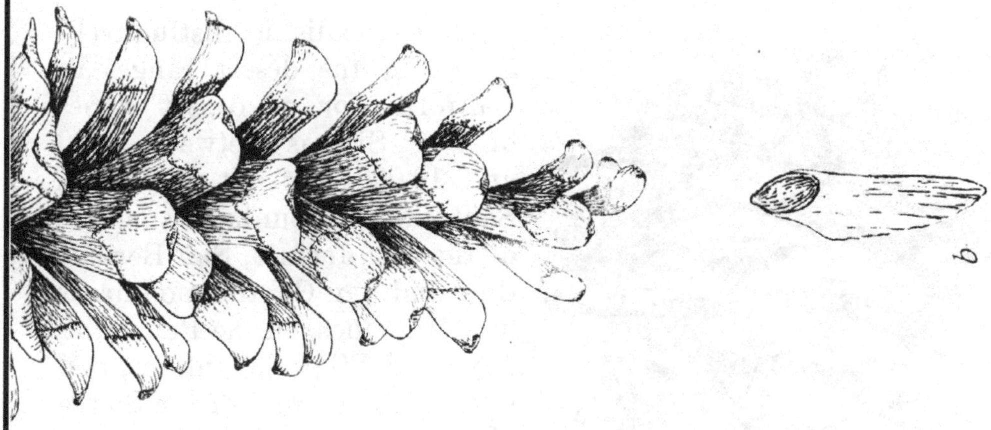

Fig. 7. SUGAR PINE (*Pinus lambertiana* Dougl.). *a*, Open cone with stalk, ⅔ nat. size; *b*, seed, nat. size.

Fig. 8. SILVER PINE (*Pinus monticola* Don). *a*, Cone, ⅔ nat. size; *b*, seed, nat. size.

Fig 10 LIMBER PINE (*Pinus flexilis* James). Open cone nat size

Pinus flexilis is distinctively a tree of the desert ranges in California or of desert slopes of our Sierras between 7000 and 10,000 feet. It occurs on Santa Rosa Mountain, summits of the San Jacinto, San Bernardino and San Gabriel mountains and Mt. Pinos, on the Panamint, Inyo and White mountains, and on the eastern wall of the Sierra Nevada from Mono Pass to Monache Peak. It is widely distributed in the desert ranges of Nevada (where on account of the scarcity of timber it is highly valued as "White Pine") and extends east to the Rocky Mountains from Alberta to New Mexico.

Foxtail Pine

Fig. 11

Foxtail Pine, often called Balfour Pine (Pinus balfouriana Jeffrey), is a sub-alpine tree commonly 20 to 45 or rarely 55 feet high. The trunk-axis is cone-shaped, 1 to 3 feet in diameter at the base, and in old or storm-beaten trees at timber line generally projects through the crown as a dead and shining splinter point. The bark is reddish-brown, smoothish but superficially checked into square plates. The branches are stout, rather short, or irregular in length, with half-drooping branchlets thickly clothed with needles spreading equally all around the stem and thus resembling a fox's tail.

Pinus balfouriana was discovered by the botanical explorer John Jeffrey on the Scott Mountains west of Mt. Shasta and named by the collector for his friend, Professor John Hutton Balfour of Edinburgh University.

Pinus balfouriana is a local species confined to two widely separated high-montane areas, one in the North Coast Ranges, the other

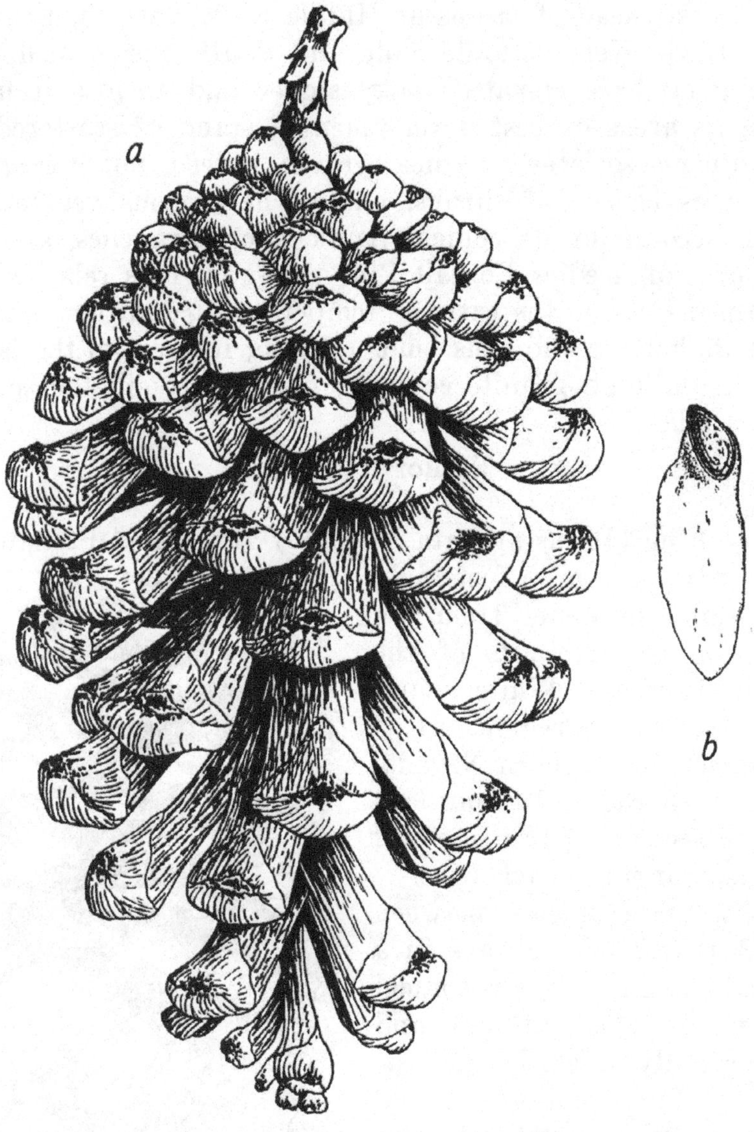

Fig 11 FOXTAIL PINE (*Pinus balfouriana* Jeffrey). *a*, Open cone; *b*, seed nat. size

in the southern Sierra Nevada. In the North Coast Ranges it occurs on the Scott Mountains (as just indicated) and on the Yollo Bollys in Tehama County. In the southern Sierra Nevada it is distributed from the head of the San Joaquin North Fork southward to Monache Peak, at altitudes of 9000 to 12,000 feet. It is abundant at Coyote Pass, on the Whitney and Chagoopah plateaus, and about Bullfrog Lake and East Vidette. It also occurs on Mt. Silliman, Alta Peaks,

Kaweah Peaks, head of basins of Middle Fork Kaweah, East Fork Kaweah, Little Kern, Middle Tule and South Fork Kern rivers.

Growing on bare elevated rocky slopes and cirques it usually forms in its areas of best development a stand of scattered trees without other associates. Typical groves, indeed, not infrequently occupy areas barren of shrubs, and of herbaceous vegetation as well. On account of its conical trunk, short branches, and short dense masses of needles Foxtail Pine shows obvious relation in its architectural form to the extreme temperature conditions and high winds of its habitat and thus lends striking interest to the isolated colonies in the high granite country, particularly in the southern Sierra Nevada.

Hickory Pine

Fig 12

Hickory Pine (Pinus aristata Engelm.) also called Foxtail Pine, is a bushy tree 15 to 55 feet high, which enters California from the eastward on the summits of the Panamint, Inyo and White mountains of the desert region. It is more common in southern Nevada, Utah and Colorado. In southern Nevada, on account of tree scarcity, it is valued for mine timbers.

Pinus aristata grows between 8000 and 12,000 feet, always in a thin stand. The trunks are conical—that is very thick at base and tapering rapidly upwards after the manner of high montane pines. A characteristic stand of this species occurs on Telescope Peak in the Panamint Range between 9000 and 11,200 feet, where the trees are 25 to 45 feet high. The writer in 1918 measured four that at four feet above the ground gave diameters of 4 to 6 feet. On the summits of the White Mountains in northeastern Inyo County between Big Prospector Meadow and

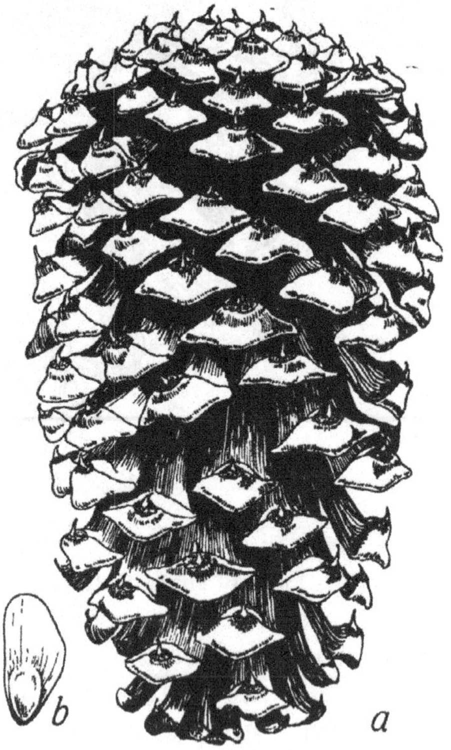

Fig 12 Hickory Pine (*Pinus aristata* Engelm) *a*, Cone, *b*, seed. nat size

the cañon of Silver Creek, we find a regular stand of mature trees that is very open but not extremely thin.

The bark of the trunk varies from light reddish brown to dark brown and is lightly fissured. The cones spread or hang from the under side of the branchlet, very frequently being borne on pendulous branchlets.

Yellow Pine

Figs 13, 14

Yellow Pine, more specifically Western Yellow Pine (Pinus ponderosa Dougl.), is a forest tree of the first class, 60 to 225 feet

Fig 13 YELLOW PINE (*Pinus ponderosa* Dougl) Open cone nat size

high with long narrow open crown and trunk 2 to 8 feet in diameter. The branches in mature trees are horizontal or even drooping, the lower ones in forest stands regularly deciduous upwards, giving clear shafts 40 to 100 feet in length. Isolated trees bear very long branches nearly to the ground. The bark in typical trees is 2 to 4 inches thick, tawny or yellow-brown, divided by fissures into large smoothish or scaly-surfaced plates 1 to 4 feet long and ¼ to 1½ feet wide. Other trees and especially younger trees, or trees of scattered colonies outside the main Yellow Pine belt, have dark red-brown bark or black bark, being respectively the indefinite "Bull Pine" or "Black Pine" of woodsmen.

Pinus ponderosa grows on fertile moist mountain slopes and plateaus, dry or rocky ridges, granite cliffs, high fertile valleys, low gravelly valleys and arid desert slopes. It not only has an extensive distribution but it also grows in a greater variety of habitats and is subject to greater ranges of temperature and precipitation than any other North American tree. It is of our Californian trees more widely distributed horizontally and vertically than any other species, it is the species most abundant in individuals, and it is (forestrally considered) of greatest commercial importance.

It generally forms the major portion of the stand in the Yellow Pine belt of the Sierra Nevada, that is between 1500 and 5000 feet at the north, to 3000 to 6000 feet in the central portion, and 5000 to 7000 feet at the south. Its most common associates are Black Oak, Incense Cedar, Sugar Pine and White Fir. It is common on all the high ranges of Southern California such as the San Gabriel, San Bernardino, San Jacinto, Palomar, Santa Aña and Cuyamaca mountains. It occurs in the Santa Inez, San Rafael, and Big Pine mountains. In the South Coast Ranges it has been found in the Santa Lucia, Santa Cruz and Mt. Hamilton ranges but is absent from the Mt. Diablo and San Carlos ranges. In the North Coast Ranges it is widely distributed, especially at higher altitudes. It is also abundant in and around Mt. Shasta and extends northeasterly over the plateaus and ranges of the Modoc lava bed country. Beyond our borders it is found throughout the Rocky Mountains and in the intermediate region north to British Columbia and south to northern Mexico and Lower California.

Fig 14.
Yellow Pine
seed. nat. size.

The wood is light or heavy, fine- and straight-grained and usually very resinous; it is pale yellow, reddish yellow or sometimes very light in color. Certain trees manufactured into lumber give planks which are practically indistinguishable from Sugar Pine and are graded in the yards with Sugar Pine stock and sold as such. "Apple Pine," which has a fragrant wood, is one of the lumberman's high-grade varieties of Yellow Pine.

The Yellow Pine is, as said above, the most abundant and widely distributed tree of California and is particularly characteristic of the Sierra Nevada, where it attains its finest development. The largest trees most commonly grow along the ridges and it is the ridges which the trails ordinarily follow. Here the traveler may journey day after day, over needle-carpeted or grassy ground, mostly free of underbrush, amidst great clean shafts 40 to 150 feet high, of really massive proportions but giving a sense of lightness by reason of their color, symmetry, and great height. No two trunks in detail of bark are modeled exactly alike, for each has its own particular finish; so it is that the eye never wearies of the fascination of the Yellow Pine but travels contentedly from trunk to trunk and wanders satisfyingly up and down their splendid columns—the finest of any pine.

Jeffrey Pine

Fig 15

Jeffrey Pine (Pinus ponderosa var. jeffreyi Vasey) is a forest tree 60 to 120 or 170 feet high, typically with rusty or wine-colored trunks, the bark broken into roughish plates. It is ordinarily distinguished from the Yellow Pine by its larger cones which are 5 to 8 inches long. When open the cones are much denser and shaped like an old-fashioned straw hive. The prickle of the umbo is often more slender. The seeds are often obovate, 5 to 7 lines long, with a wing 12 or 13 lines long.

In its extreme form it is very distinct from the typical form of the Yellow Pine. On the other hand the transition forms to the Yellow Pine are quite as numerous as the individuals of the true Jeffrey Pine and occupy as extensive an area. It is chiefly by reason of this consideration that Jeffrey Pine is here retained in varietal rank. The following peculiarities in this connection may, however, be set down. While working as a forester in the Eagle Lake district of Lassen County, Mr. Vance S. Brown has observed that newly exposed trunk bark scales in Jeffrey Pine are wine-colored or reddish-purple, in Yellow Pine they are a distinct yellow. It may also be noted that the pine beetle, Dendroctonus jeffreyi Hopkins, is found only on Jeffrey Pine, while Dendroctonus brevicomis LeConte works on Yellow Pine, although not peculiar to the latter species.

Fig 15 Jeffrey Pine (*Pinus ponderosa* var *jeffreyi* Vasey) *a*, Cone, b, seed ½ nat size

In its typical form Jeffrey Pine is found at elevations of 6000 to 9000 feet. It inhabits the San Jacinto and San Bernardino mountains, high Sierra Nevada (common on western but especially abundant on eastern slope), Mt. Shasta, Siskiyou and Scott mountains and the Yollo Bolly Range. Growing at higher altitudes than the Yellow Pine it forms thinner forests, is more often flat-topped or broken, and has a greater trunk diameter relatively to its height.

Jeffrey Pine was first collected by John Jeffrey, Scottish botanical explorer, in Shasta Valley, October 24, 1852. He disappeared after leaving San Diego in 1853 and doubtless lost his life on the Colorado Desert in Southern California.

Beach Pine

Fig 16

The Beach Pine (Pinus contorta Dougl.) is commonly a littoral tree with dark green crown 10 to 25 feet high. The trunk is ¼ to 1¼ feet in diameter and clothed in dark roughly fissured thick bark. Typically the trees are dwarfed or with a very irregular crown owing to their exposed situation. This form is common on the Mendocino bluffs and on the sandhills at Samoa near Eureka. At Crescent City about one mile easterly from the town is a small grove, the trees 20 to 30 feet high and forming a pure stand. On the Mendocino White Plains occurs a dwarf cane-like form 2 to 5 feet high and bearing freely very slender cones.

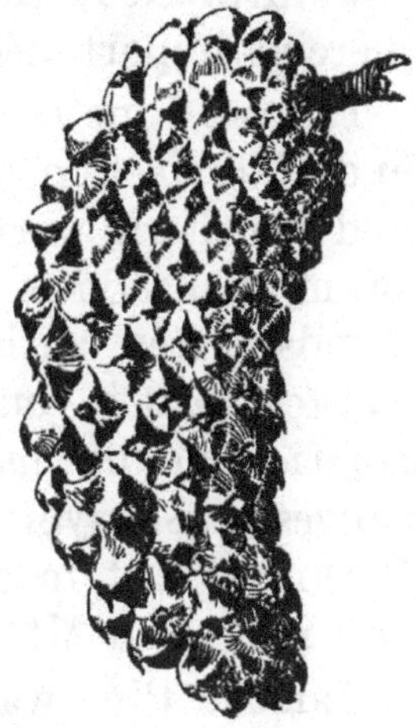

Fig 16 Beach Pine (*Pinus contorta* Dougl), cone nat size.

Pinus contorta ranges from Point Arena to the sand-dunes of the Oregon and Washington coasts and northward to Alaska, its altitudinal range being from sea-level to 500 feet. In the Cascades of Oregon and Washington it passes into the var. murrayana Engelm.

Of slight economic value for its wood, since the trunks are small and faulty, it is of great importance in the natural reclamation of the sand-dunes on the coast of Oregon and Washington, where it forms depressed trees whose lower limbs lie prostrate as a protection to the uneasy sands. The cones are borne prolifically, every tree being full of them. To the traveler the Beach Pine is always an interesting feature of the north coast, because the form of the tree is so frequently an expression or record of its struggle with the elements, and while of insignificant stature it often attains a dignified age, its maximum longevity being two hundred and seventy-five and possibly three hundred years.

Tamrac Pine

Fig 89

Tamrac Pine, the Lodgepole Pine of the Northwest (Pinus contorta var. murrayana Engelm.), is a forest tree with symmetrical and rather dense crown, 50 to 80 or rarely 125 feet high. Sometimes it appears at timber line, as on Mt. San Jacinto and through the Sierra Nevada from Mt. Whitney north to Mt. Shasta; then it is dwarfed or storm-battered, 10 to 30 feet high, or even occurring semi-prostrate. Its bark is light gray, remarkably thin, usually ¼ inch thick, very smooth but flaking off into thin scales. The wood is fine- or coarse-grained, reddish brown and hard. While little utilized in the past in California it will eventually be of commercial importance.

Tamrac Pine occurs throughout the Sierra Nevada between 5000 and 7000 feet in the north and 7000 to 11,000 feet in the central and southern part. It is an especially characteristic feature of swampy meadows or moist mountain slopes where it forms dense stands, often without admixture of other species. In Southern California it is found on the San Jacinto, San Bernardino and San Gabriel mountains; in northern California it grows on the high ranges of Siskiyou County and Modoc County. Beyond our borders it ranges east to the Rocky Mountains of Colorado and Montana and recurs on Mt. San Pedro Martir in Lower California.

Tamrac Pine was discovered on the Siskiyou Mountains October 21, 1852, by John Jeffrey. It was named Pinus murrayana in honor of Andrew Murray of Edinburgh, Scotland, member of the Botanical Association which sent Jeffrey to the Pacific Coast.

Big-cone Pine

Figs. 1, 17

Big-cone Pine, often called Coulter Pine (Pinus coulteri Don), is a tree 40 to 70 feet high with rather dense conical or more often spreading crown with usually long lower branches. The foliage is yellowish green. The trunk is 1 to 2½ feet in diameter; the bark is dark, roughly broken so as to form an irregular network of longitudinal fissures and sometimes loosening superficially into long thinnish scales.

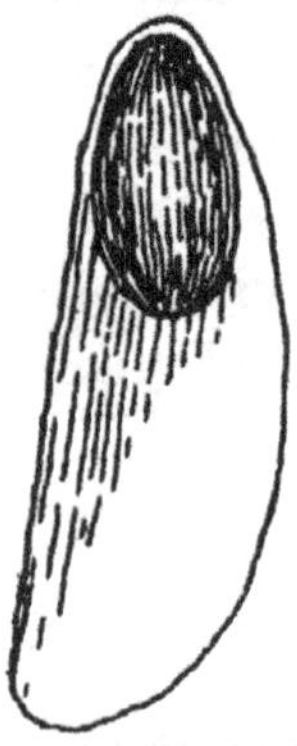

Fig. 17. *a*, Big-cone Pine seed, nat. size.

Pinus coulteri grows on dry or rocky mountain slopes at 2500 to 6000 feet, chiefly on the lower margin of or below the Yellow Pine belt wherever it occurs in Yellow Pine region. It is most abundant on the San Bernardino and San Gabriel mountains, ranging south to the San Jacinto, Palomar, Santa Aña, Cuyamaca, Balkan and Laguna mountains, and into Lower

Fig 18. Digger Pine (*Pinus sabiniana* Dougl.). Open cone, ⅔ nat. size.

California. Northward it occurs on the Santa Inez, San Rafael, Santa Lucia, Gabilan, San Carlos and Mt. Hamilton ranges. On Mt. Diablo, the most northerly locality, a few trees occur about Mitchell Rock on the north side of the mountain near the village of Clayton at 800 feet altitude.

While having a general resemblance to young Yellow Pine, Big-cone Pine is a very different tree and is easily recognized by its heavier masses of foliage, stout twigs, and its great cones with their eagle's-claw appendages to the scales.

As a timber tree it is of slight importance. On account of its scattered or at least thin growth the shafts retain their lower limbs and do not form good clear logs. Moreover, the wood is very resinous, very brittle, warps with great facility and in consequence is never used for lumber except in those localities where good timber is scarce. In Southern California the tree is of most economic importance as a ground cover on arid slopes where other species will not grow. It is one of the "nut pines," but the nuts were either not harvested by the native tribes of the region or only when they could not procure the nuts of the One-leaf Piñon, which they much preferred.

Digger Pine

Figs 18, 19, 20

Digger Pine (Pinus sabiniana Dougl.) is a singular tree 40 to 50 or occasionally 90 feet high with a very open crown and thin gray foliage. The trunk is 1 to 4 feet in diameter, frequently slanting, commonly branching at 5 to 15 feet from the ground into a cluster of slender erect branches which form a broom-like top. Rarely one sees a tree in which the trunk-axis bears only lateral branches and persists through the crown. The bark is dark gray, roughly furrowed.

Fig 19 Digger Pine seed, nat size.

Pinus sabiniana grows in dry hot foothills and sometimes in gravelly valleys, chiefly between 50 and 2000 feet altitude, although frequently reaching 5000 feet at the south. It always occurs as a scattered growth, often by itself or if with other trees most commonly with Blue Oak. In the Coast Ranges it is widely although not continuously distributed, occurring in the Gabilan, San Carlos, Mt. Hamilton, Mt Diablo, Napa, Vaca and Mayacmas mountains. It does not grow in the summer fog belt of the North Coast Ranges (Redwood region), although it reaches the coast in the Santa Lucia Mountains (South Coast Ranges) and is scattered along the east slope of the Santa Cruz Mountains. In the Sierra Nevada it is the most characteristic tree of the foothills where it is usually the only pine except at its upper limits. It ranges south to Tehachapi and the Sierra Liebre, north to the Sacramento River cañon and the cañon of the South Fork of Salmon River near Bennet.

Digger Pine was discovered by David Douglas, during his travels in California in 1830 and 1831. He himself described the species as new, and named it Pinus sabiniana in honor of Joseph Sabine, Secretary of the London Horticultural Society, the organization which sent the Scotch explorer to California. Douglas dated his paper February 4, 1831, at Mission San Juan Bautista, the old Spanish village under the east base of the Gabilan Range where this species still grows plentifully.

Pinus sabiniana is one of the "nut pines." The "nuts" or seeds were a favorite article of food with the native tribes and they had no more useful tree save the oaks. It is small wonder then that the

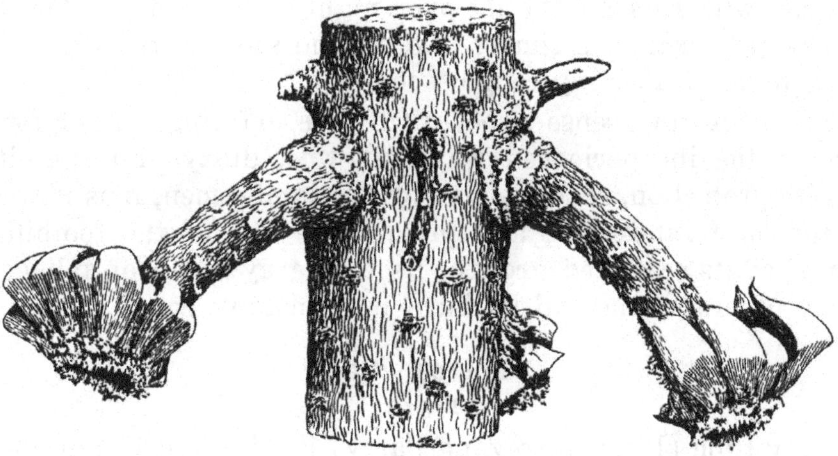

Fig 20 Digger Pine Branch with whorl of three cone-stalks, the stalks bearing the persistent scales which parted from base of cone when the cones fell

Indian looked on in distress whenever the "white man" cut down a Digger Pine. The green cones were, also, an article of food with native tribes, who removed the soft core in June and ate it uncooked; they also ate the young buds and the inner bark in the spring, according to Powers, when food was scarce.

Digger Pine is the white man's familiar name for the species and its origin is clear to those who know that "Digger" is his opprobious and impartial designation of all native tribesmen. The name Squaw Pine is less commonly used and has a similar origin, but pays tribute to the women who harvested the winter's store of nuts. Gray Pine, Gray-leaf Pine, and Blue Pine are names which interpret the hue of the foliage. Bull Pine is also in local use, a name indiscriminately applied by woodsmen to any economically inferior or non-yielding species or local variety, while Hooked Bull Pine is a variant of this.

The wood of the Digger Pine is coarse-grained, not strong and exceedingly pitchy; when seasoned it is very hard and flinty. It has been largely used for fuel, makes a remarkably hot fire, and is consumed in quantity by the hoisting engines at quartz mines in the Sierra Nevada. Sometimes it is used for mine props in the stopes, but rots very quickly when wet, rarely lasting more than six months. In regions where timber is scarce it has been utilized in construction, although it makes an unsatisfactory lumber when sawn, on account of its faculty for warping. The woodsmen say with humorous exaggeration "Boards from the mill stacked outside to season will walk off the lot over night." Like the Yellow Pine it varies remarkably in timber quality and selected trees have been known to yield a strong tough wood.

Scarcely in any sense a beautiful tree, offering no comfort of shade to the inexperienced wayfarer who, dusty and sun-bitten, seeks its protection, scorned, too, by the lumbermen, it is nevertheless the most interesting and picturesque tree of the foothills on account of its scattered growth, its thin gray cloud of foliage, its variety of branching and its burden of massive cones.

Torrey Pine

Fig. 21

Torrey Pine (Pinus torreyana Parry) is a low crooked or sprawling tree 15 to 35 feet high, or sometimes straight and 60 feet high. It is local on the San Diego coast about Del Mar near the mouth of the Soledad River, extending southward toward San Diego about 8 miles and inland about 1½ miles. It also occurs on Santa Rosa Island. It is remarkable for its peculiar cones, its very long needles and especially for its restricted habitat which has attracted to it a

Fig. 21. TORREY PINE (*Pinus torreyana* Parry). *a*, Cone; *b*, seed. nat. size.

great deal of attention. The trees themselves are, however, disappointing. They are insignificant in stature and habit, and notwithstanding that they are the only trees where they grow they dominate the landscape so little as scarcely to be noticed except by the traveler acquainted with the peculiarities which give them a singular interest.

The trees growing in the San Diego County locality, the first to be made known, were discovered in 1850 by Doctor C. C. Parry, botanist of the Mexican Boundary Survey, and named in honor of Professor John Torrey of Columbia College, long a student of western plants and a visitor to California before the days of the Overland Railroad.

Parry Piñon

Fig. 22

The group of true nut pines includes four forms which have passed as species, namely Mexican Piñon (Pinus cembroides Zucc.) of Mexico with 4 needles in the sheath; Parry Piñon (Pinus parryana Engelm.) with 2, 3, or 4 needles; Rocky Mountain Piñon (Pinus edulis Engelm.) with 2 or 3 needles; and One-leaf Piñon (Pinus monophylla Torr.) with one needle. These various pines do not differ from each other appreciably in habit or in reproductive organs; their cones are remarkably similar; their seeds are destitute of wings. They are currently distinguished by the number of needles in the sheath, but sheaths with 1, 2, or 3 needles may occasionally be found on a single individual of Pinus monophylla as well as Pinus edulis, while the number of needles in Pinus parryana is notoriously variable. Unless more positive differences can be established it is necessary to reduce our forms to varieties of the Mexican Piñon, Pinus cembroides.

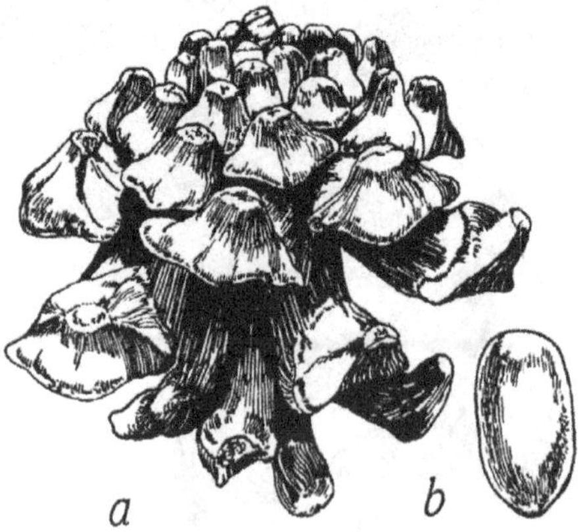

Fig 22 Parry Piñon (*Pinus cembroides* var. *parryana* Voss) *a*, Cone; *b*, seed. nat size

Parry Piñon, or Four-leaf Piñon (Pinus cembroides var. parryana Voss), is a short-trunked low tree 15 to 30 feet high. It inhabits the dry slopes of the San Jacinto and Cuyamaca mountains and extends southward in Lower California to Mt. San Pedro Martir. It occurs sparingly in Southern California and is known to me only from the following stations: Jim Hill; mesa at west base of El Toro; slopes near Coyote Cañon; near Julian; Larkin Station.

One-leaf Piñon

Fig 23

One-leaf Piñon (Pinus cembroides var. monophylla Voss), the "Nut Pine" of the Nevada ranges, is most commonly a low round-headed tree with very short trunk, remotely suggestive of an

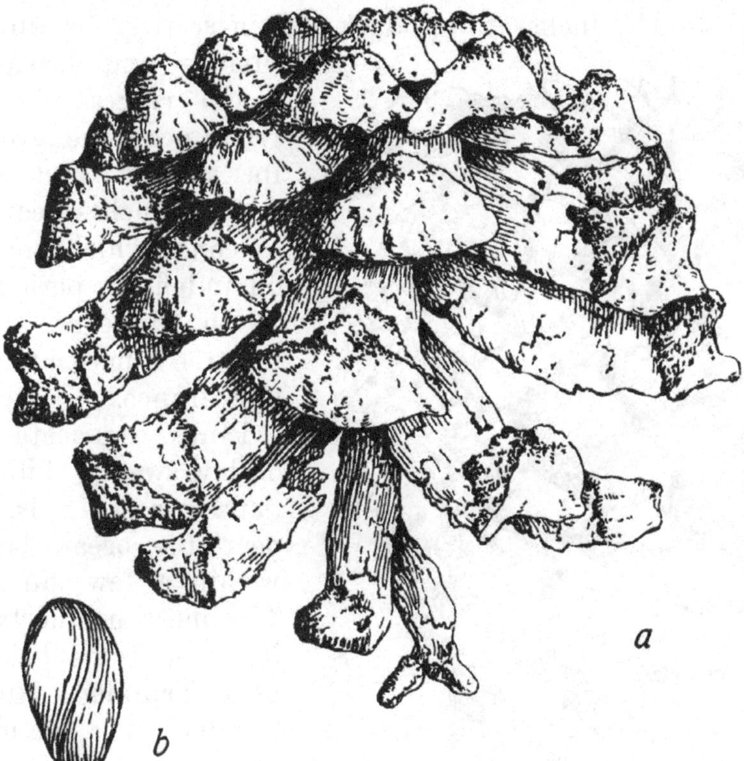

Fig. 23. One-leaf Pinon (*Pinus cembroides* var *monophylla* Voss). *a,* Open cone; *b,* seed nat size

old apple tree, 8 to 20 or sometimes 45 feet high. It grows on arid slopes or rocky walls and is distributed through the desert regions of Utah, Nevada and Arizona westward to the desert ranges of California (White, Panamint and Providence mountains), eastern wall of the Sierra Nevada, Tehachapi, San Emigdio, and San Rafael mountains, thence along the desert slopes of the San Gabriel, San Bernardino and San Jacinto ranges, and so on south into Lower California. On the west slope of the Sierra Nevada it occurs in a few circumscribed localities, in Piute Cañon near Pate Valley (Grand Cañon of the Tuolumne River), Kings River, along the west wall of the Kern Cañon and southward into the lower Kern River country.

Bishop Pine

Fig 24

Bishop Pine (Pinus muricata Don) is a tree 40 to 80 feet high with roundish or flat crown and trunk 1 to 3 feet in diameter. The bark is 1 to $1\frac{1}{2}$ inches thick, dark red in section, brown on the surface and broken into rough ridges. The trunk from near the ground to its summit and all the main branches bear circles of cones which persist for an indefinite period, often 15 to 25 years, and give the tree a most remarkable appearance.

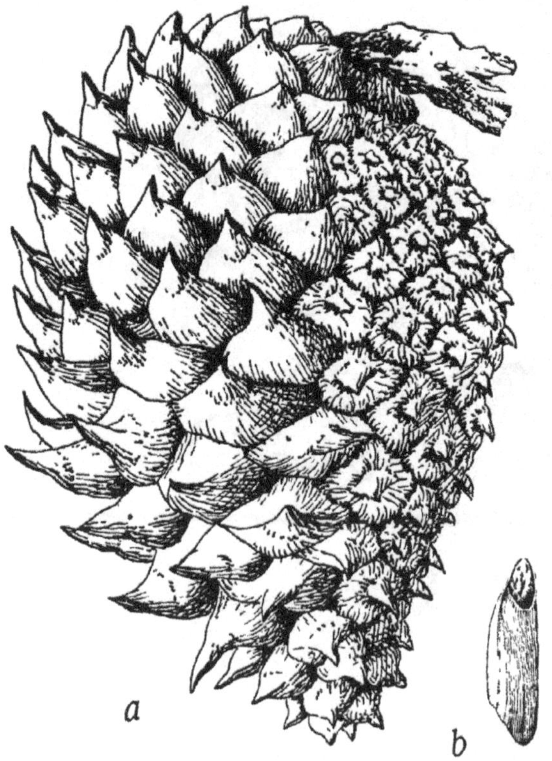

Fig 24 Bishop Pine (*Pinus muricata* Don) *a*, Closed cone, *b*, seed

Pinus muricata grows on low swampy hills, moist flats or rocky hills, always near the ocean beach or within a few miles of it. The most northerly locality is on Luffenholz Creek near Trinidad, Humboldt County. It is abundant on the plain of the Mendocino coast (Inglenook to Ft. Bragg) and extends southward to the Sonoma coast where it reaches its greatest development, forming pure groves of small extent on the low swampy hills. It also occurs on the rocky and clay hills of the Point Reyes Peninsula on the eastward slope of the Inverness Ridge from near Tomales Point southward nearly to Bolinas Bay. South of the Golden Gate it occurs on Huckleberry Hill at Monterey and near San Luis Obispo where it was originally discovered in 1830 by Doctor Thomas Coulter, a botanical traveler. The original station suggested the common name, Bishop Pine, but the scale tips thickened at the tip like a bishop's cap also emphasize the happiness of the vernacular appellation. Southward still it recurs in the region of Mission La Purissima.

The wood of Bishop Pine is very resinous, light, hard and rather coarse-grained. It is sometimes used for piling as the light-brown heartwood is very durable. The tree is at present of most economic importance as a wind-break.

Monterey Pine

Figs 25, 26

Monterey Pine (Pinus radiata Don) is a beautiful symmetrical tree, or in age with flattened or broken top, 30 to 70 or 115 feet high. The foliage is a rich dark green. The trunk is 1 to 4 feet in diameter and clothed with a roughly fissured bark which is hard and more nearly black than that of any other Californian pine.

Pinus radiata grows on dry rocky or sandy hills near the sea and is confined to a few limited localities, namely, Pescadero, Monterey and San Simeon, all on the Californian mainland, and Santa Rosa, Santa Cruz and Guadalupe islands off the south coast. It is abundant locally in Monterey where it covers with a dark green forest the low range of hills half encircling the town, extends southward to Carmel Bay, and in scattered colonies to Malpaso Creek a few miles south of Pt. Lobos.

Fig 25 Monterey Pine, closed cone small type nat size

Fig. 26. Monterey Pine (*Pinus radiata* Don). *a*, Open cone; *b*, seed nat. size.

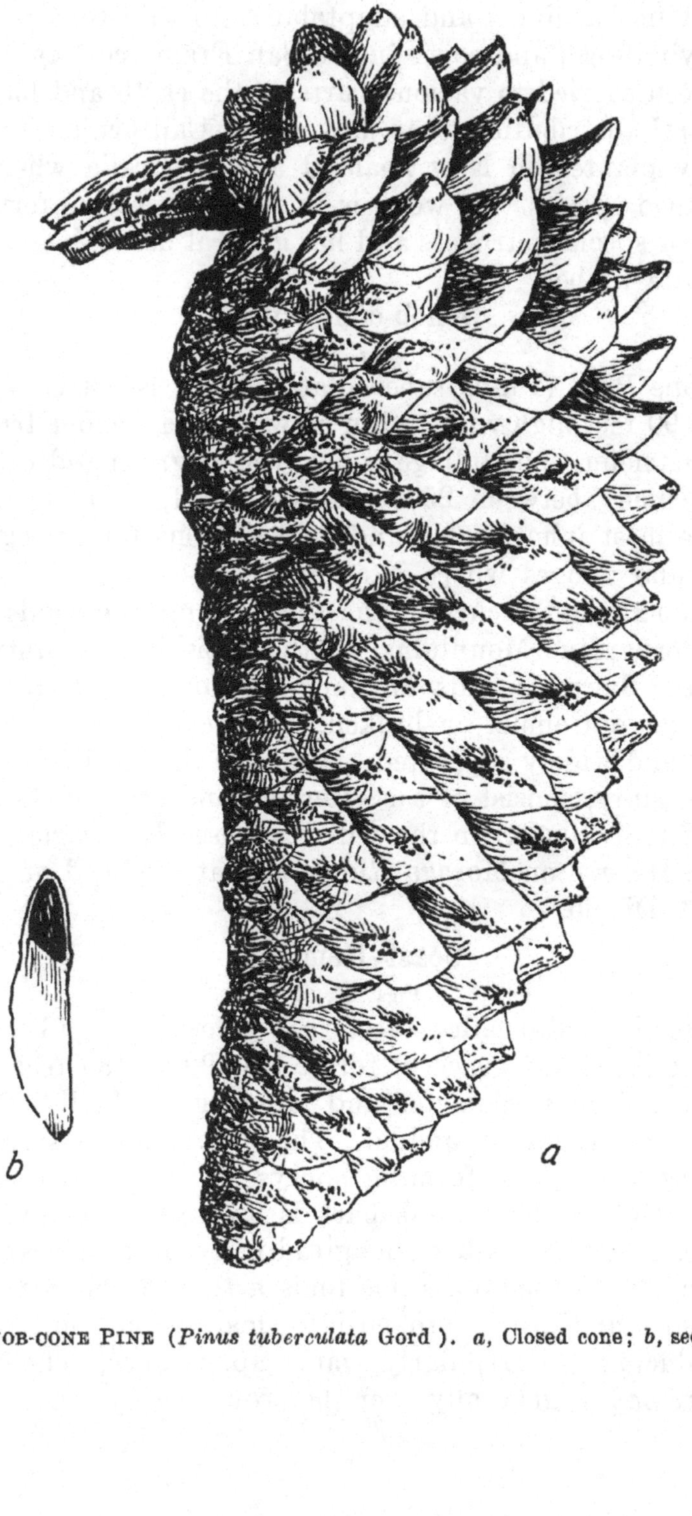

Fig 27. KNOB-CONE PINE (*Pinus tuberculata* Gord). *a,* Closed cone; *b,* seed nat. size.

It grows rapidly, attains its maturity in 30 to 50 years, and on account of its hardiness and adaptability to thin soils it has been used for windbreak and cover in the San Francisco Bay region. It has also been carried to various parts of the earth and has a wider horticultural distribution than any other Californian tree. It is extensively planted in New Zealand and Australia where it is a highly valued species. Its wood is in some respects inferior but it is prized for special purposes and has general use as a second-grade or substitute timber.

Knob-cone Pine

Fig. 27

Knob-cone Pine (Pinus tuberculata Gord.) is a tree 5 to 30 or sometimes 90 feet high with a thin crown and a slender trunk ⅓ to 1 foot in diameter. The foliage is pale yellow-green and rather thin. It grows chiefly between 2500 and 4000 feet, on rocky slopes or ridges, the most inhospitable and arid stations for tree growth in the particular ranges where it occurs.

Pinus tuberculata occurs in the Coast Ranges, extends south to the San Bernardino Mountains, north to Siskiyou County, thence south in the Sierra Nevada to Mariposa County. Its range is exceedingly discontinuous or broken, the localities being comparatively few and widely separated. Moreover the localities, except in Siskiyou County, consist of a scattered stand of a relatively small number of individuals. In the San Francisco Bay region it occurs on Mt. St. Helena, on Moraga Ridge, and at the head of Sycamore Cañon, Mt. Diablo.

Coast Hemlock

Figs 28, 29

Coast Hemlock, also called Western Hemlock (Tsuga heterophylla Sarg.), is a large but graceful forest tree 90 to 180 feet high, the trunk 1 to 4 feet in diameter and tapering gradually, the crown narrow or sometimes pyramidal. The branches are slender, with finely hairy branchlets, forming sprays which droop cascade-wise but not pendulous. The trunk bark is brown on the surface, dark red inside, shallowly fissured longitudinally or nearly smooth, ⅓ to ¾ inch thick. Sometimes one finds a tree in which the bark is twice as thick and deeply broken into small oblong plates one inch high, producing an irregularly warty appearance. The cones are borne more or less generally over the crown.

COAST HEMLOCK

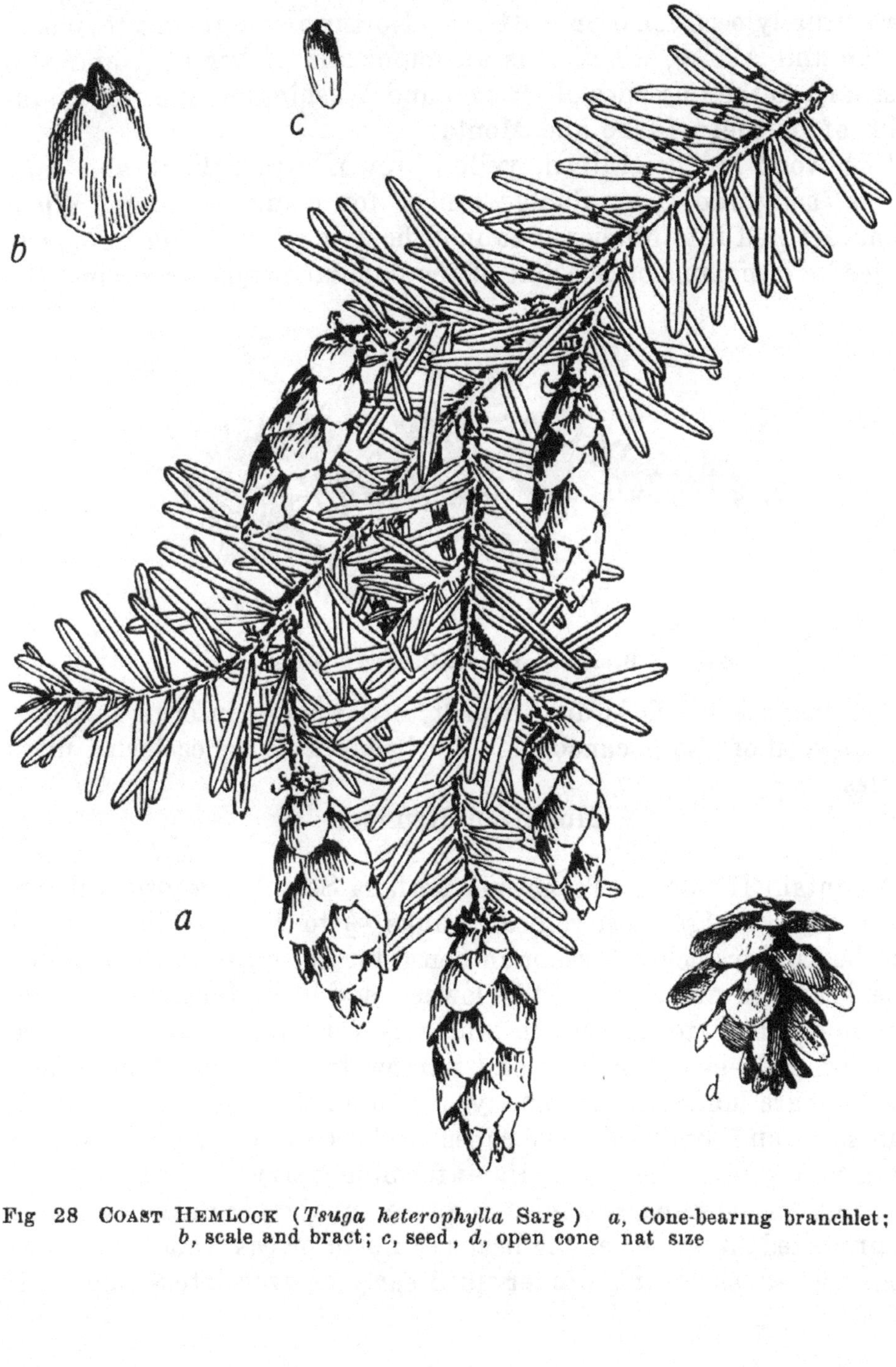

Fig 28 Coast Hemlock (*Tsuga heterophylla* Sarg) *a*, Cone-bearing branchlet; *b*, scale and bract; *c*, seed, *d*, open cone nat size

Tsuga heterophylla grows in the immediate vicinity of the coast from the Gualala River, Sonoma County, to Humboldt and Del Norte counties as a very subordinate associate of the Redwood, the trees usually occasional or scattered. Northward it ranges to Washington and Alaska, where it is an important timber tree, and also eastward to the Cascades of Oregon and Washington and the mountains of northern Idaho and Montana.

The wood is fine-grained, yellow-brown, rather light and soft, works easily and is a valuable timber for many purposes. When manufactured into lumber it cannot, however, be sold on its merits under its own name on account of the popular prejudice against the

Fig. 29. Branchlet of Coast Hemlock ⅓ nat size.

name hemlock which is due in part to the many inferior qualities of the wood of Tsuga canadensis, the hemlock of the eastern United States.

Mountain Hemlock

Fig. 30

Mountain Hemlock (Tsuga mertensiana Sarg.) is a graceful tree 20 to 90 feet high with conical trunks ½ to 2½ feet in diameter and bearing branches (except in dense forest) quite to the ground. The branches are slender, the lower ones long, forming a broad pyramidal base to the crown which is very promptly contracted upward and ends in a long and narrow top. The branches above the base are horizontal or mostly drooping, the branchlets slender, pubescent and drooping. The cones are borne in the top of the tree, on drooping branchlets, sometimes forming heavy clusters.

Tsuga mertensiana, a timber-line tree, inhabits high slopes chiefly in protected situations at the heads of north or east cañons in moist places where snowbanks linger until early or even late summer. It

Fig 30. **Mountain Hemlock** (*Tsuga mertensiana* Sarg.). *a*, Cone-bearing branchlet; *b*, scale and bract; *c*, seed. nat. size, the seed a little enlarged.

usually occurs in small pure somewhat open groves or clusters of limited extent. This subalpine type of tree, the most characteristic, retains its lowermost branches and is readily recognized by the habit of its crown, pyramidal at base but narrowed above, with drooping branchlets and pendulous whip-like leader. The crowns are usually dense throughout and sometimes remarkably slender above, presenting columns of foliage 15 to 30 feet high and sometimes not exceeding 2 feet in diameter except at the broad base.

In winters of heavy snowfall in the Sierra Nevada the heads of small saplings are bent over to the ground. Such deep snow does not always go off completely during the following summer but the position of these little trees may, sometimes, be betrayed by the protrusion of a bow-like trunk through the surface of the drift.

At somewhat lower altitudes or in protected stations Mountain Hemlock forms large-sized forest trees either in pure stands or in association with Silver Pine, Red Fir or Tamrac Pine. A tree near the base of Mt. Lyell, measured by the writer in 1909, was 80 feet high and 5½ feet in trunk diameter at 4 feet. On the east wall of Matterhorn Cañon is a pure stand of large-sized trees with trunks naked towards the ground, the branches horizontal in middle of crown and up to the tip. One tree measured 55 feet high and 5 feet in trunk diameter. It was typical of the larger sized trees throughout this fine grove.

Tsuga mertensiana is found in the Sierra Nevada between 8000 to 11,000 feet at the south and 6000 to 10,000 feet at the north. The southernmost locality is on Bubbs Creek, South Fork Kings River. Northward it is found on Glass Mountain, Goosenest Mountain and Mt. Shasta, thence west to Trinity Mountains, Marble Mountain, and the Siskiyou Mountains. Beyond our borders it ranges far north to Alaska and east to Montana.

Tideland Spruce

Fig 31

Tideland Spruce (Picea sitchensis Carr.) is a handsome forest tree 75 to 180 feet high with conical crown, wide-spreading rigid branches and drooping branchlets. The trunk is 3 to 20 feet in diameter at the base where it flares most remarkably in older trees; at 6 or 8 feet above the ground the trunk diameter may be only half that at the base. The trunk bark is reddish brown, developing

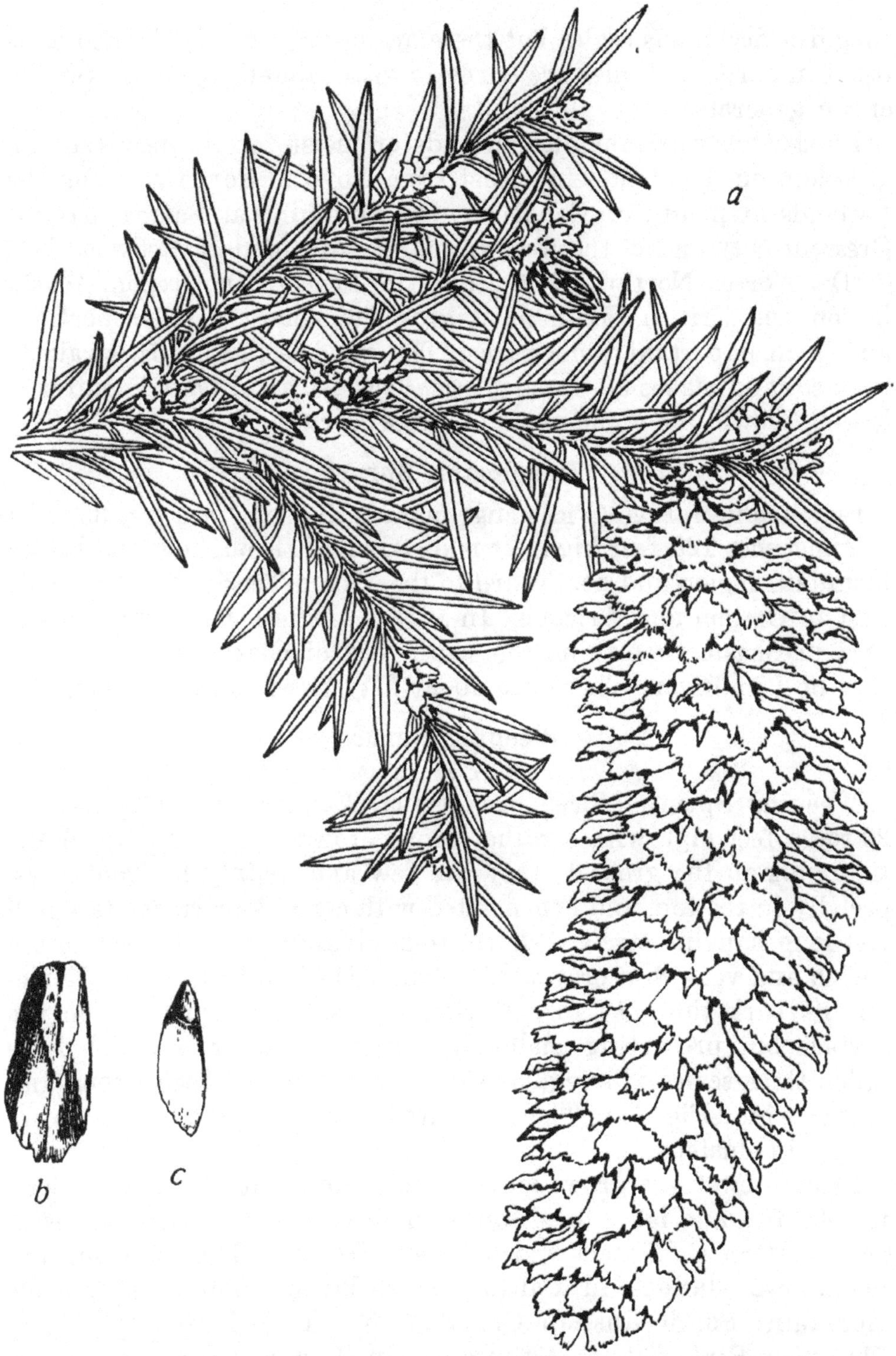

Fig 31. TIDELAND SPRUCE (*Picea sitchensis* Carr) *a*, Cone-bearing branch, *b*, scale and bract, *c*, seed nat. size.

roughish deciduous scales but these are not so sharply defined as is usual in spruces. Cones are borne in great abundance and over the crown generally.

Picea sitchensis inhabits lowlands or moist forests near the sea. It occurs on the Mendocino coast from Noyo to Fort Bragg, on the lowlands at mouth of Eel River, the sand-hills at Samoa, flats at Crescent City and in the western margin of the main Redwood belt of Del Norte. Northward it ranges to Alaska. In Oregon, Washington and British Columbia it grows to almost vast proportions and is an important timber tree. The wood is light, soft, straight-grained, and makes an excellent saw-log. As a cultivated conifer it is well known under the name of Sitka Spruce.

Engelmann Spruce

Engelmann Spruce (Picea engelmannii Engelm.) is a tree 30 to 75 or sometimes 125 feet high. It is distributed throughout the Rocky Mountain region and northward to the Yukon Territory. It extends west to Oregon and Arizona. In 1922 a single locality was discovered in Shasta County on the south fork of Clark Creek, a tributary of Pitt River. There are about fifty trees at this station.

Weeping Spruce

Figs. 32, 33

Weeping Spruce (Picea breweriana Wats.) is a subalpine tree 20 to 95 feet high with a rather broad crown. The branches clothe the trunk to the ground; they are few and mainly horizontal, especially in the top, and ornamented with cord-like branchlets 1 to 6 feet long hanging straight down, thus giving a formal effect to the stiffish and very thin crown. The trunk is ½ to 3½ feet in diameter, its bark thin (½ inch thick), whitish and smoothish on the surface but presenting shallowly concave scars from which have fallen thick scales of irregular shape, mostly 1 to 4 inches long and half as wide (Fig. 33). The cones are borne in the top of the crown, mostly in clusters.

Picea breweriana grows on cool moist slopes at altitudes of 6000 to 8000 feet and is confined to the summits of a few high ranges in northwestern California and adjacent Oregon. The following are the known stations in California: Siskiyou Mountains; Marble Mountain; South Russian Creek on North Fork Salmon River; Thompson Peak, Salmon Mountains. In Oregon it is reported from

FIG. 32. WEEPING SPRUCE (*Picea breweriana* Wats). *a,* Branchlet with open cone; *b,* scale and bract. nat. size.

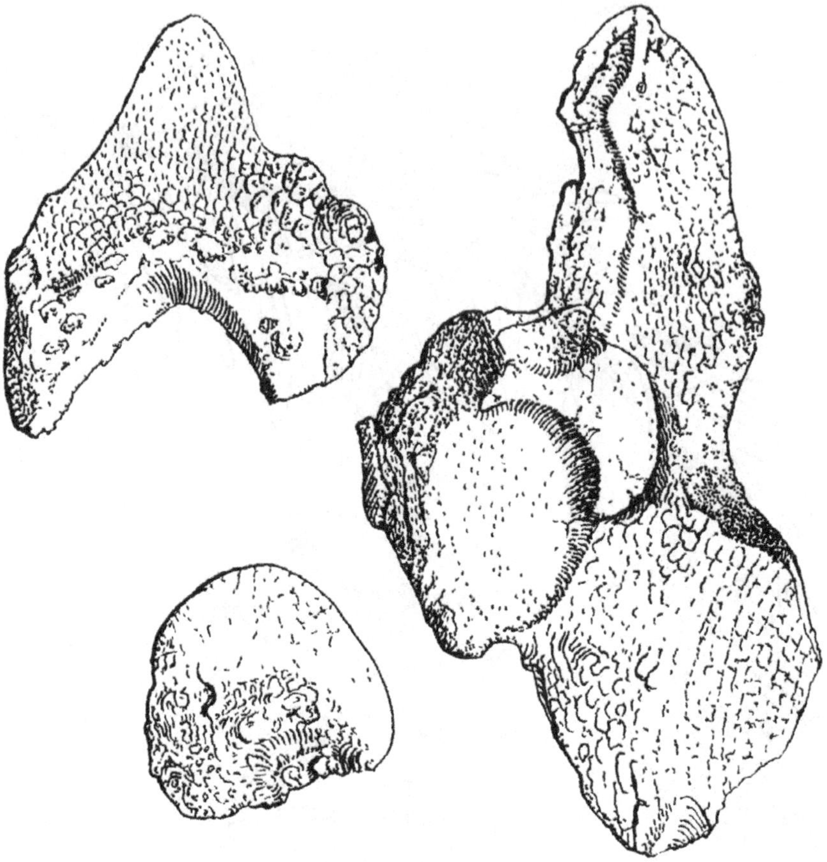

Fig 33. Weeping Spruce (*Picea breweriana* Wats) Scales which have fallen from the trunk bark nat size

the Oregon Coast Range, Chetco Range, and high mountain tops south of Rogue River. Near Marble Mountain the writer measured a tree 95 feet high and 10 feet 10 inches in circumference at 4 feet above the ground.

Associated with Mountain Hemlock, Silver Pine and Red Fir, Weeping Spruce is a most remarkable species inhabiting the tops of mountain ranges lying in a country long scourged by fire. The trees are found only at the highest altitudes, usually at the heads of north cañons where even in July or September one may find a lingering snowbank which feeds their roots with water. The appearance of the trees is so singularly different from that of any other conifer that they cannot ever be mistaken and at once arrest the attention of the traveler.

Douglas Fir

Fig 34

Douglas Fir (Pseudotsuga taxifolia Britt.) is often called Douglas Spruce. It is a magnificent forest tree 70 to 250 feet in height, in dense stands exhibiting clear trunks 100 to 150 feet high and 4 to 8 feet in diameter. The bark on young trees is smooth, gray, or mottled, on older trunks 1 to 6½ inches in thickness, soft or putty-like, dark brown, fissured into broad heavy furrows, in cross-section showing alternate layers of red and white. The branchlets are usually drooping.

Pseudotsuga taxifolia inhabits fertile mountain slopes, moist cañons, dry gravelly valleys and rocky ridges and with us favors north or east slopes. It is found from sea-level to 4000 feet, nearly throughout the North Coast Ranges, associated with Redwood in the Redwood belt and with the Tan Oak in the middle ranges; in the inner ranges it is less common or altogether absent. In the South Coast Ranges it is found in the Santa Cruz and Santa Lucia mountains. In the Sierra Nevada it occurs chiefly between 2500 and 6000 feet and extends as far south as the San Joaquin River. Beyond our borders it ranges north to British Columbia and through the Rocky Mountains to northern Mexico.

In Washington and Oregon where it reaches its most splendid development it occurs in great abundance and furnishes the most important and widely used structural timber in western America under the trade name of "Oregon Pine." No other conifer yields wood of such lightness, strength, flexibility and durability. On account of its large size, timbers for bridges, big buildings and ship construction can be had ranging from the smaller grades of scantlings to sticks 1 or 2 feet square and up to 100 or 200 feet in length. Incredible as it may seem some of these large sizes are regularly quoted in trade lists.

By reason of the hardiness of the tree, its rapid growth, strong reproductive power and adaptability to a wide range of conditions, it is, in view of its wood value, an unequaled species forestrally for its region and destined to play a large part in scientific forest operations of the future, especially in the Pacific Northwest.

Douglas Fir is the type species of the genus Pseudotsuga which consists of two species. Pseudotsuga is most nearly related to the Spruces from which it differs in its peculiar bracts and pollen-sacs.

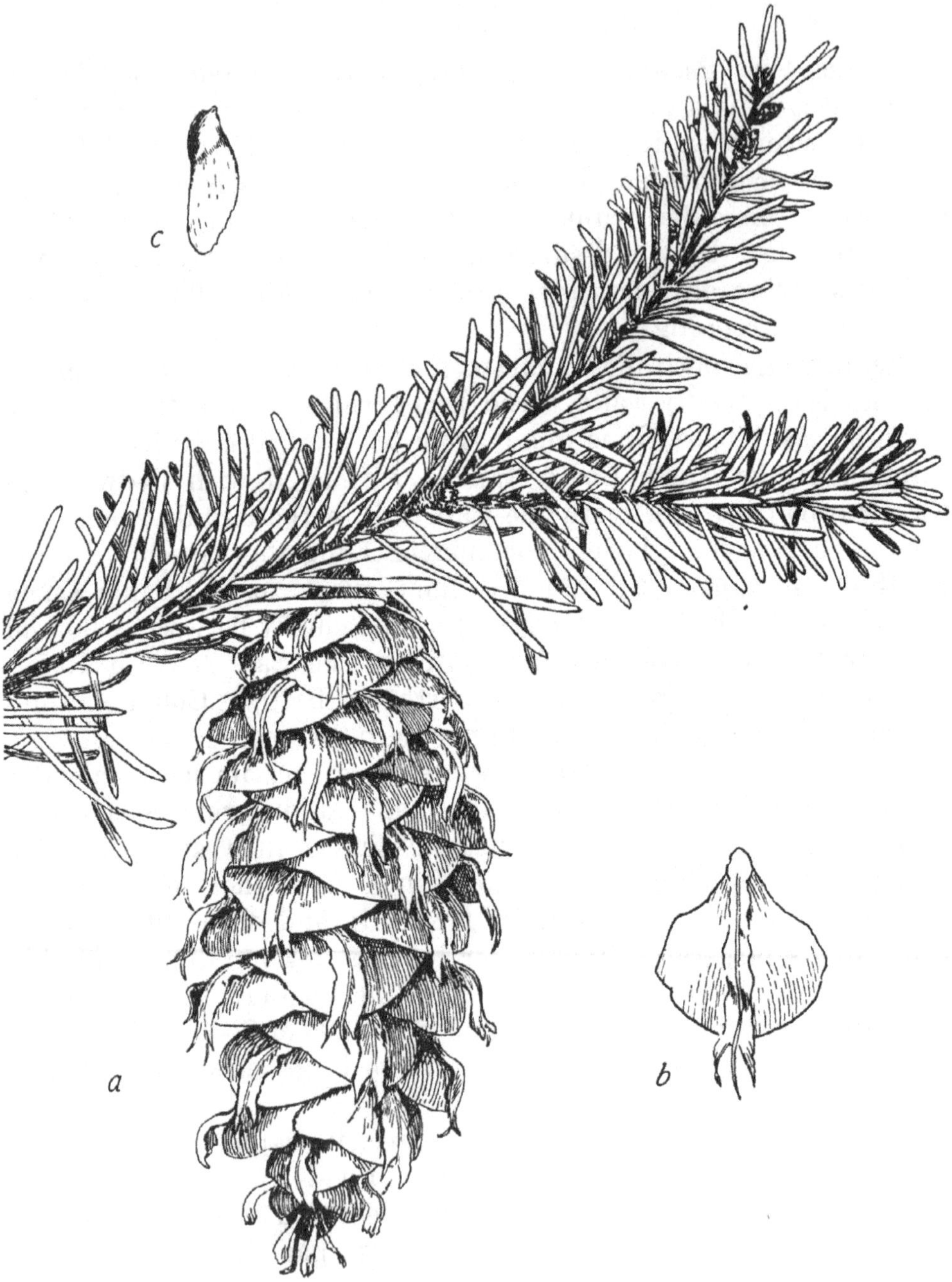

FIG. 34. DOUGLAS FIR (*Pseudotsuga taxifolia* Britt). *a*, Cone-bearing branchlet; *b*, scale and bract; *c*, seed. nat. size.

It differs from the Hemlocks in its peculiar cones and pollen-sacs; the resin vesicles of Hemlock seeds are not found in Pseudotsuga seeds. From the Firs this genus differs very greatly in wood, foliage and character of cones. Fir cones, moreover, are borne erect and fall to pieces on the tree. The status of Pseudotsuga is more exactly indicated in the following tabulation:

RELATIONSHIP OF PSEUDOTSUGA.

Spruce-like characters	*Hemlock-like characters*
General habit, character of branching.	Leaves petioled, blunt.
Pendulous branchlets.	Leader of young trees often pendent.
Leaves spreading all around stem.	Cones pendent.
Cones pendent, borne all over tree, of much the same shape and size as those of spruces.	Cone-scales persistent.
Cone-scales persistent.	*Fir-like characters*
Seeds without resin vesicles.	Bark thick, roughly fissured, not scaly as in spruces.
	Resin pockets in young bark.
	Bracts exserted.

Big-cone Spruce

Big-cone Spruce (Pseudotsuga macrocarpa Mayr) is a tree 30 to 90 feet tall with a broad pyramidal crown and very long lower branches. The bark is dark or black. In most respects this species is very similar to its near relative, the Douglas Spruce or Douglas Fir. It grows on the sides of sheltered cañons or ravines or on cool north slopes at altitudes just below the Yellow Pine (mainly between 3000 and 5000 feet) and forms small groves or colonies, usually growing by itself. It is distributed from the San Emigdio Range westward to the San Rafael and Santa Inez ranges, and south to the San Gabriel, San Bernardino, San Jacinto, Santa Aña, Palomar and Cuyamaca mountains. It also recurs on Mt. San Pedro Martir in Lower California. Its wood is fine-grained, tough and hard, but yields a coarse lumber remarkably inferior to that of its high-class relative, Douglas Fir.

White Fir

Fig 35

White Fir (Abies concolor L. & G.) is a forest tree 60 to 150 or 200 feet high with a long narrow crown composed of flat sprays declined or spreading horizontally and with a trunk 1 to 8 feet in diameter and clear of limbs for 30 to 100 feet. The trunk bark is smooth, silvery or whitish on young trees; on old trees it is 2 to 4 inches thick, broken into rounded ridges separated by heavy fissures,

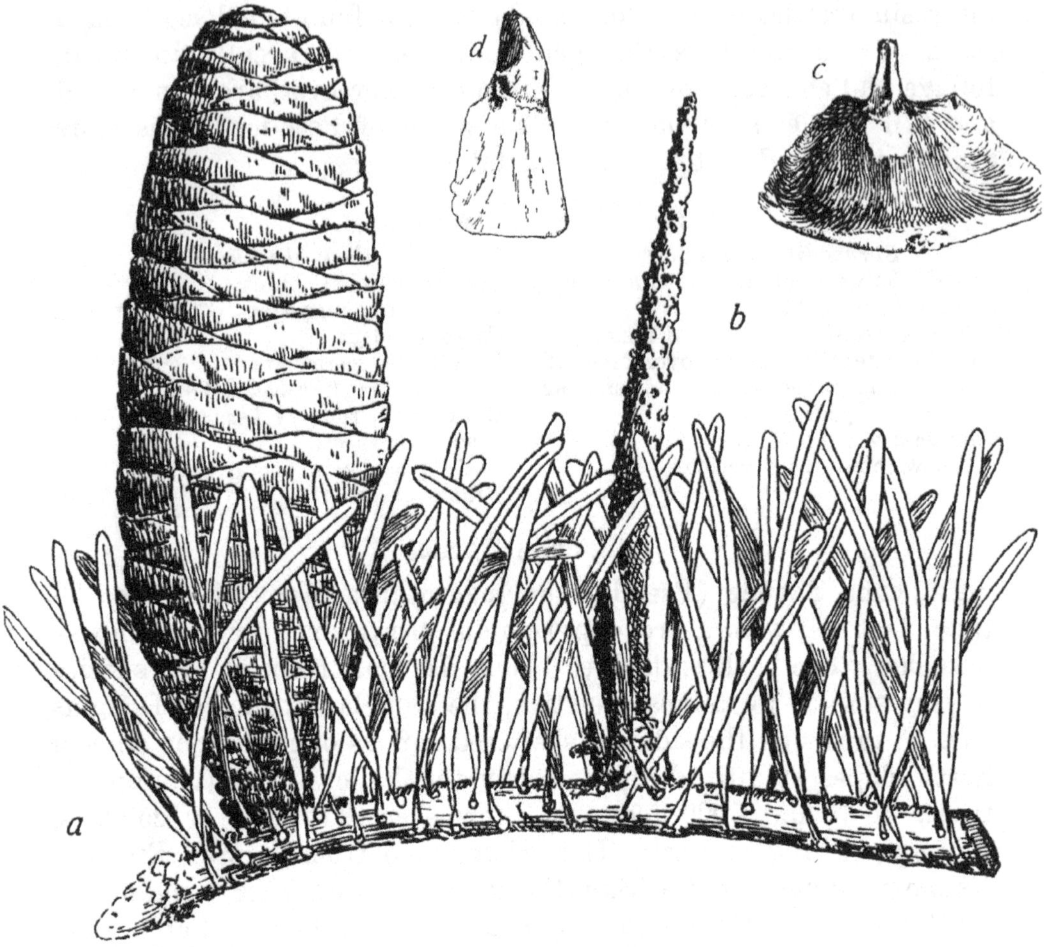

FIG 35 WHITE FIR (*Abies concolor* Lindl & Gord). fruiting branch *a*, Cone; *b*, axis from which scales have fallen; *c*, scale and bract, *d*, seed nat size

gray or drab-brown, in section showing dull-brown areas separated by a coarse light-colored mesh.

Abies concolor inhabits fertile mountain slopes, rocky ridges or plateaus or cañon walls. Associated with Yellow Pine, Sugar Pine and Incense Cedar, it is one of the four most important forest trees in the main timber belt of the Sierra Nevada where it grows between 2500 and 7500 feet at the north and 5000 to 8000 feet at the south. In Southern California it occurs from about 5000 to 9000 or even 11,500 feet on all the high coastal ridges from Mt. Pinos south to the San Gabriel, San Bernardino, San Jacinto, Volcan and

Cuyamaca mountains. It recurs on Mt. San Pedro Martir in Lower California.

White Fir is of rapid growth and aggressively reproductive since it will establish seedlings in its own territory wherever there is sufficient moisture. The wood is soft, light, coarse-grained, fairly strong and useful for timber, but not used as yet to any great extent except as small lumber for fruit boxes. As a source of box lumber it is of particular importance to the California fruit industry.

Single individuals of exceptional size are often found on gentle mountain slopes or in cañon bottoms in deep soil. Near the upper end of Lake Merced (east of Yosemite) is a tree 160 feet high with trunk 8½ feet in diameter at 4½ feet above the ground. On the Cuddihy trail to Dutch Henrys on the Klamath River, the writer measured a tree 24 feet in circumference at 4 feet. These individuals of exceptional size which arrest the attention of the traveler are in most cases apparently of great age.

Lowland Fir

Lowland Fir (Abies grandis Lindl.) is a forest tree 40 to 160 or even 250 feet high with narrow conical crown of horizontal branches and trunks 1½ to 4 feet in diameter. The trunk bark is white or light brown, smooth or shallowly broken into low flat ridges; in section the inner bark is light brown, the outer dark red with a mesh of purple lines through it. The foliage is dark green and shining; on the lower horizontal branches the leaves spread by a twist at base in two opposite ranks and so make a flat spray, those originating on top of the stem having the peculiarity of being much shorter than those coming from the sides.

Abies grandis inhabits low hills or valleys and is distributed along the coast from northern Sonoma County to Del Norte County and far north to southern British Columbia, thence to Montana. In California it is a very subordinate species. It is associated with Tideland Spruce on the Mendocino bluffs and occurs as scattered trees among the dominant Redwoods to a distance of 15 or 20 miles from the coast, attaining its greatest size with us in the Redwood belt of Del Norte County.

Its wood is light, soft, fairly strong and is milled in California on a small scale. The lumber is put to rough or temporary pur-

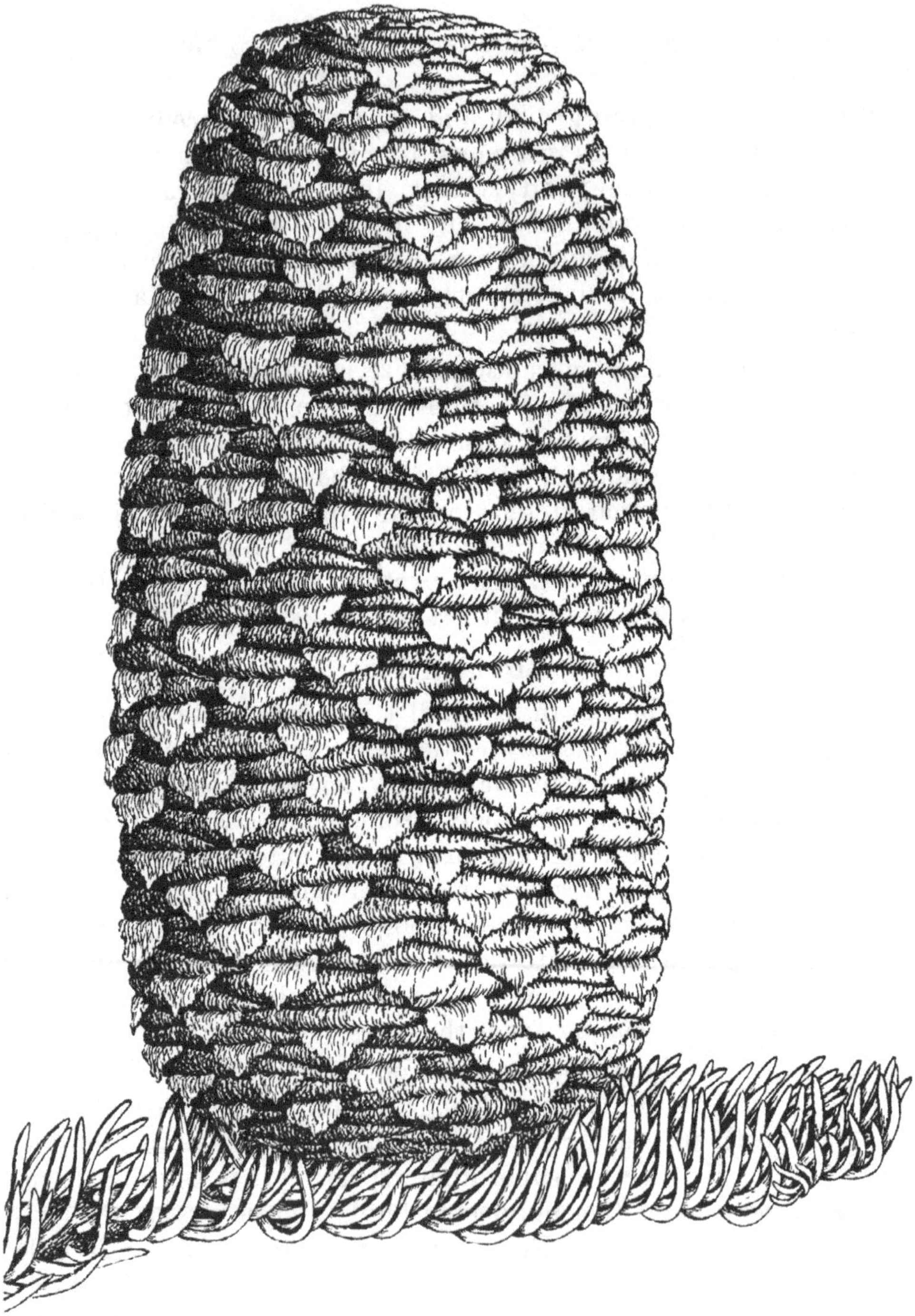

Fig. 36 Red Fir (*Abies magnifica* Murr), the form called "Shasta Fir" with exserted bracts. The ordinary form is similar save that the bracts are concealed. nat. size.

poses. It rots very quickly in contact with soil but is held in local esteem along the north coast for shelving in dry goods stores and for similar purposes where an odorous wood is a desideratum.

Red Fir

Figs. 36, 37

Red Fir (Abies magnifica Murr.) is a forest tree of great beauty, 60 to 175 feet high, with a narrowly conical crown composed of numerous horizontal or declined strata of fan-shaped sprays and a trunk 1½ to 5 feet in diameter. The bark on young trees is whitish or silvery, on old trunks dark red, very deeply and roughly fissured, in section showing reddish brown areas set off by a sharply defined purple mesh.

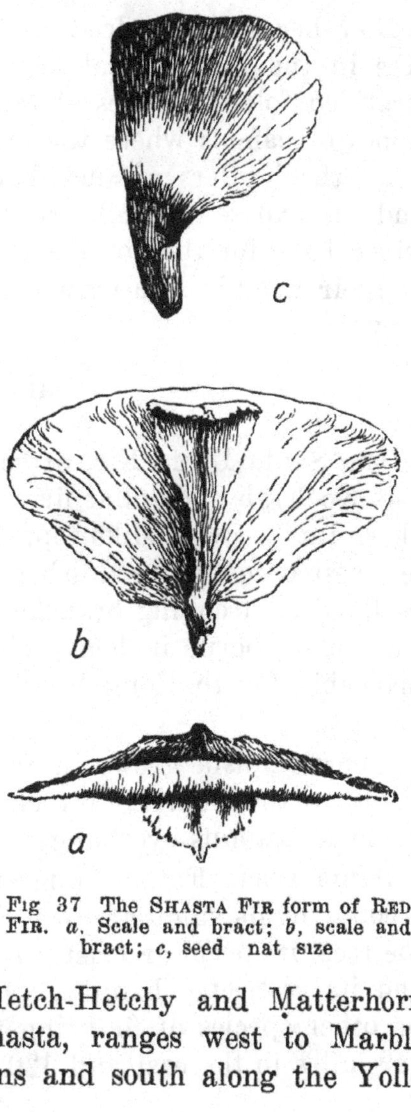

Fig 37 The SHASTA FIR form of RED FIR. *a*, Scale and bract; *b*, scale and bract; *c*, seed nat size

Abies magnifica grows on moist slopes or about swampy meadows, on rocky ridges, cliffs or granite plateaus. In the Sierra Nevada, where it attains its greatest development and is chiefly, though not exclusively, confined to the west slope, it is one of the more important species in the upper portion of the main timber belt, occurring mainly between 5000 and 7000 feet at the north and 6000 and 8500 feet at the south. It occurs as far south as the Greenhorn Range (south limits). Fine trees grow on the high plateaus in the Kaweah Peaks region. Perhaps the finest, most extensive and nearly pure forest of Red Fir covers Rancheria Mountain north of the Tuolumne River between Hetch-Hetchy and Matterhorn Cañon. It is abundant on Mt. Shasta, ranges west to Marble Mountain and the Trinity Mountains and south along the Yollo

Bolly range to Snow Mountain in Lake County. Northward it extends into southern Oregon.

Shasta Fir (var. shastensis Lemm.) is a form of the species with thicker and somewhat shorter cones and exserted bracts. It is not otherwise different, grows with the species on Mt. Shasta, and also in the North Coast Ranges and southern Sierra Nevada.

The wood of Red Fir is heavy, soft, strong, straight- and fine-grained and with a reddish tinge. On account of its durability and the large size of the saw logs it is valued in bridge building and for shaft timbers in the mines along the Mother Lode.

No other fir and indeed no other conifer of California equals this tree in the symmetrical beauty of its crown and its dark green stratified foliage. Trees of remarkable symmetry inhabit protected slopes or cañons where the moisture conditions are favorable. Beyond the borders of such habitats, especially at higher altitudes and on exposed granite ridges, grow isolated or scattered trees which have for the botanical traveler an equal interest on account of their wind-broken crowns and the irregularity of the resulting growths.

Santa Lucia Fir

Fig 38

The Santa Lucia Fir, often called Bristle-cone Fir (Abies venusta Koch), is a singular fir with a narrow crown abruptly tapering above into a steeple-like top. The trunk is ½ to 2½ feet in diameter, vested in light reddish brown bark and bearing short slender declined or drooping branches often nearly or quite to the ground. The cones, borne in heavy clusters in the top of the tree, are remarkable for the long bristles which protrude from between the scales.

Abies venusta grows in cañons or on sheltered slopes or sometimes on the summit of rocky ridges. It is confined to the Santa Lucia Mountains overhanging the Monterey coast.

Santa Lucia Fir is the most remarkable fir tree in the world on account of its singular form, its sharp-pointed leaves alike all over the tree, its peculiar bristly cones, the small number of individuals and its restricted habitat. Moreover it is isolated geographically, no other species of fir being found within 225 miles to the north, 149 miles to the east and 120 miles northeasterly.

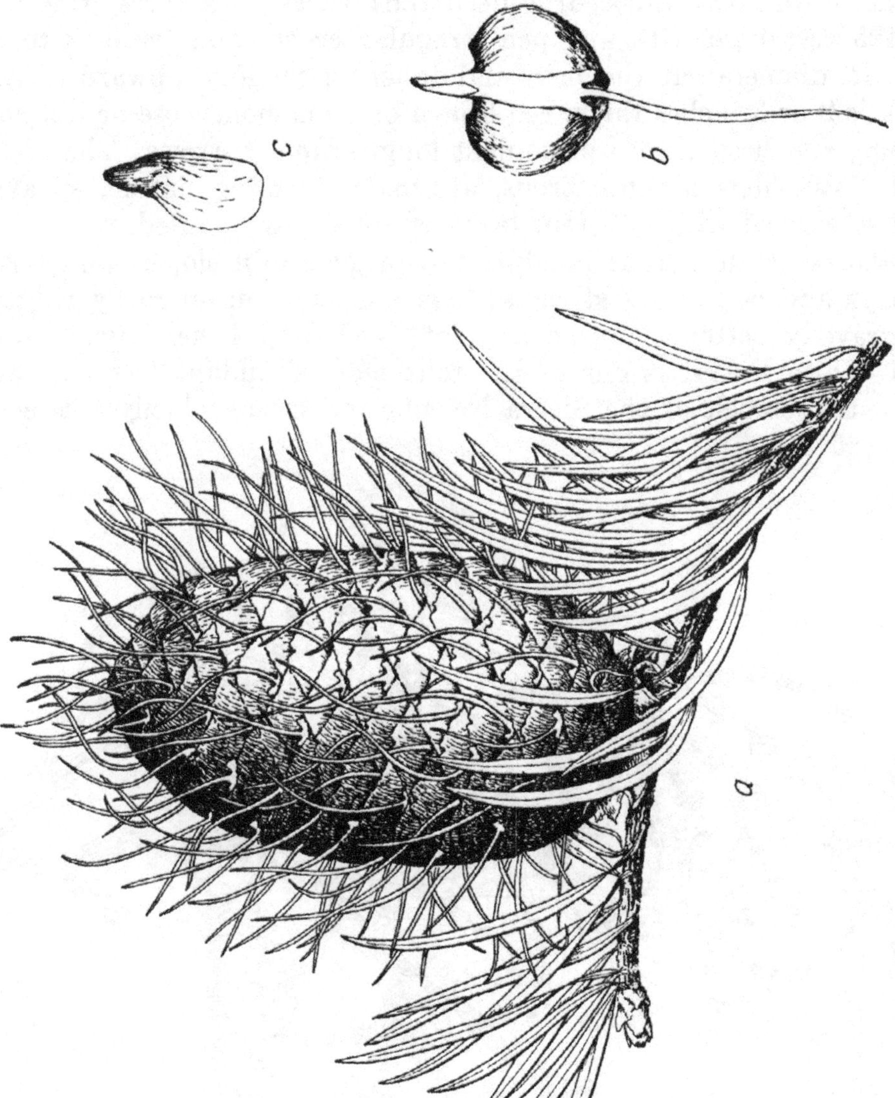

Fig 38. Santa Lucia Fir (*Abies venusta* Koch), remarkable for its long sharp-pointed leaves and long bristly bracts *a*, Cone-bearing branchlet; *b*, scale and bract; *c*, seed nat size

Incense Cedar

Fig. 39

Incense Cedar (Libocedrus decurrens Torr.) is a forest tree 50 to 125 feet high with an open irregular crown and trunk 2 to 7 feet in diameter at the base and tapering rapidly upward. The bark is 2 or 3 inches thick, red-brown or cinnamon, loose or fibrous in age, and broken into prominent longitudinal furrows. The ultimate branchlets are numerous, alternate, forming flattish sprays and so clothed with adherent leaves as to appear jointed.

Libocedrus decurrens inhabits fertile mountain slopes, plateaus, valleys and borders of streams; it is less common on rocky ridges or gravelly bottoms. As an associate of Yellow Pine, Sugar Pine and White Fir, it is one of the four most abundant trees in the main timber belt of the Sierra Nevada and occurs chiefly between

Fig. 39. INCENSE CEDAR (*Libocedrus decurrens* Torr) *a*, Cone-bearing spray, nat. size; *b*, branchlet showing detail of leaves, 3 times nat size; *c*, seed, 1½ times nat. size.

2000 and 5000 feet at the north, 3000 to 5500 feet in the central part and 3000 to 7000 feet in the south. In the North Coast Ranges it occurs locally but generally in the high mountains east of the fog belt from Marble Mountain and Trinity Summit east to Mt. Shasta and extends as far south as the neighborhood of Mt. St. Helena. In the South Coast Ranges it occurs on the Mt. Hamilton, Santa Lucia and San Carlos ranges. In Southern California it is distributed from the San Rafael Mountains to the San Gabriel, San Bernardino and San Jacinto mountains and south to the Volcan, Palomar and Cuyamaca mountains. Beyond our borders it extends into southern Oregon, western Nevada and Lower California.

Its wood is pale and reddish-brown, soft, light, fine and straight-grained. It is exceedingly durable either in contact with soil or water and meets the local requirements for posts and telephone poles.

Canoe Cedar

Fig. 40

Canoe Cedar (Thuja plicata Don) is usually a giant tree 75 to 200 feet high with pyramidal or roundish crown, the branches long, the branchlets slender and drooping. The trunk is enormously swollen at the base, giving diameters of 4 to 16 feet at the ground but at ten feet above diminishing so rapidly as to be only about one-half the diameter at the ground. The cinnamon bark is very thin, only ½ to 1 inch thick. The branchlets are repeatedly 2-ranked, forming flat sprays thickly clothed with minute leaves.

Thuja plicata inhabits moist slopes or especially gulches, flats or river bottoms. It ranges from the Bear River Mountains in Humboldt County northward to southeastern Alaska and is abundant and of great size on the Oregon and Washington coasts. In California the trees are small and grow in a few localities of limited extent.

The wood is highly aromatic, reddish-brown, light and soft. Its most remarkable quality is durability which in connection with the size of the clear logs makes it especially suitable for manufacture into shingles for which it is extensively used. The Indians of the Northwest Coast hewed their long war canoes out of a single log, wove the fibrous bark into clothing and mats, and made dwellings and household utensils out of the wood.

Fig 40 Canoe Cedar (*Thuja plicata* Don) *a*, Cone-bearing spray, nat size; *b*, winged seed, 1¾ times nat size.

Port Orford Cedar

Port Orford Cedar, the Lawson Cypress of the gardens (Chamaecyparis lawsoniana Parl.), is a forest tree 80 to 175 feet high with narrow crown and horizontal or drooping branches ending in broad flat drooping fern-like sprays. The trunk has a tall straight shaft, its bark brown or somewhat reddish, smooth on young trees, later parting on the surface into large loose thin shreds and finally in adult trees fissured longitudinally with the furrows continuous and separated by flat ridges.

Chamaecyparis lawsoniana inhabits sandy ridges near the coast, moist slopes in the mountains or the bottoms of cool cañons or gulches. It reaches its best development on the west slope of the Oregon Coast Range between Coos Bay and Rogue River within 3 to 15 miles of the ocean. It ranges south through Del Norte and Siskiyou counties to Mad River, Humboldt County, California, and eastward to the Sacramento River Cañon. In California its localities are few.

Its wood is aromatic, yellowish white, light, fine-grained, hard and strong. It is very durable, works easily, takes a very superior finish and is highly valued for cabinet work and linen closets. Forestrally the tree is very valuable but the area in which it occurs in commercial quantity is so restricted that the supply of this timber can be depended upon to last but a very limited time.

Monterey Cypress

Fig 41a

The Monterey Cypress (Cupressus macrocarpa Hartw.) is a tree 15 to 60 feet high and grows only in two limited localities, Cypress Point and Point Lobos—the two headlands of the ocean shore at the mouth of the Carmel River near Monterey. The Cypress Point Grove extends along the cliffs and low bluffs from Pescadero Point to Cypress Point, a distance of two miles, reaching inland about one-eighth of a mile. The Point Lobos Grove is much smaller. The trees are scattered over the summits of these two headlands and cling to the edges of the cliffs, where on account of the erosive action of the ocean they are occasionally undermined and fall into the sea. Trees standing on the cliffs or exposed directly to the ocean exhibit much flattened or irregularly broken crowns and strongly flattened or board-like trunks or main branches.

Some of the largest trees, such as the individual known as the Octopus Tree, are amazingly contorted. In protected situations, even a few yards to the leeward of contorted specimens, this species develops a crown which is broadly conical and very symmetrical, with spreading finger-like tips to the main branches.

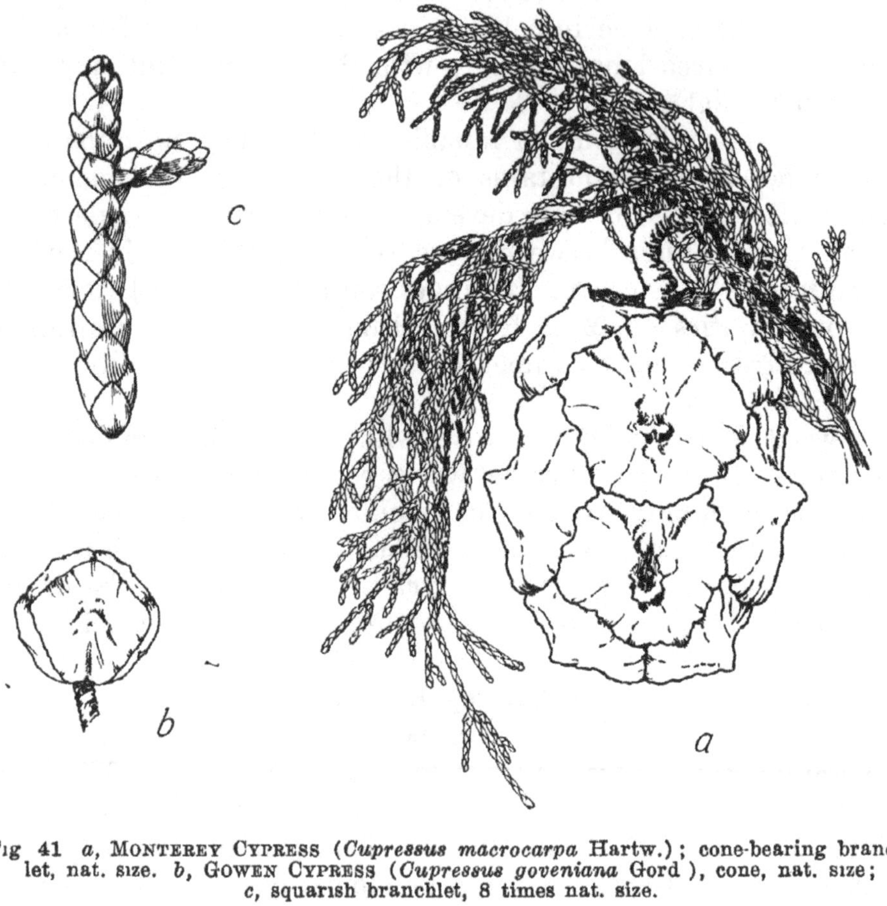

FIG 41 *a*, MONTEREY CYPRESS (*Cupressus macrocarpa* Hartw.); cone-bearing branchlet, nat. size. *b*, GOWEN CYPRESS (*Cupressus goveniana* Gord), cone, nat. size; *c*, squarish branchlet, 8 times nat. size.

The age of mature Monterey Cypress is about 50 to 300 years. There is no warrant for calling these trees one thousand or two thousand years old. Trees well-cared for in cultivation attain in twenty-five to thirty years the size and diameter of the larger trees at Cypress Point. The oldest tree whose age has been definitely determined is 284 years.

Monterey Cypress is most interesting for its remarkably restricted natural range and the exceedingly picturesque outlines characteristic of the trees growing on the ocean shore. As a result of their struggle with the violent storms from the Pacific Ocean which break on the unprotected cliffs and headlands of Cypress Point and Point Lobos, they present a variety and singularity of form which is obviously connected with their exposed habitat and lends a never-failing interest to these two narrow localities.

Of the highly picturesque trees, the most common type is that with long irregular arms. Such trees recall strikingly the classical pictures of the Cedars of Lebanon. Monterey Cypress is of course a genuine cypress and Lebanon Cedar a genuine cedar; the two do not even belong to the same family of conifers. Yet the popular story that the two are the same makes so strong an appeal to the imagination of the tourists at Monterey that the guides and promoters in the region will doubtless never cease to disseminate it. As a consequence the error goes into the daily press and the magazines and is evidently destined to flourish in perennial greenness under the guise of fact. The wide dissemination of this fiction is all the more remarkable in that in the case of all other unique features of the State, such as the Sequoias and the Yosemites, our Californians have evinced a remarkable pride in their possession without thought of inventing a duplication of them elsewhere.

Although so local a species in its natural habitat, Monterey Cypress takes most kindly to cultivation and to horticultural methods. It is widely cultivated in California for ornament, for wind-breaks and for hedges. While long-lived in coast gardens, trees planted in the dry interior valleys rarely live more than twenty-five years. As a cultivated tree Monterey Cypress has also been planted in various parts of Europe and also with especial success in Australia and New Zealand. Thus, while its natural range is narrower than that of any other member of the earth's silva known to us, its horticultural distribution is very wide and exceeds that of any other of our species except possibly the Monterey Pine.

Mendocino Cypress

On the Mendocino pine flats near the ocean grows a small slender cypress which is often cane-like and bears cones when only 1 or 2 feet high. It is a very remarkable dwarf which was named Cupressus pygmaea Sarg. It is, however, in reality a dwarf form of a broad-crowned tree 25 to 75 feet high. which is scattered at intervals through the forest of Bishop Pine on the Mendocino coast.

Gowen Cypress (Cupressus goveniana Gord.) is a compact dwarf growing almost in the center of the Monterey Pine forest at Monterey (Fig. 41 b, c).

McNab Cypress

Fig 42c, d

McNab Cypress (Cupressus macnabiana Murr.) is a shrub or small bushy tree 5 to 25 or sometimes 40 feet high with trunk ¼ to nearly 2 feet in diameter. The trunk bark is light gray and very smooth.

Cupressus macnabiana inhabits dry slopes or flats in the foothills. It is distributed in scattered stations from central Napa County to Shasta County, and recurs in the northern Sierra foothills. It is distinguished by the highly pungent and somewhat aromatic odor of the foliage, by its blue-green crowns, and by the prominent horn-like crests on the summit of the cones.

McNab Cypress was discovered by Murray and Beardsley in 1854 in Shasta County and was named in honor of James McNab, at that time Curator of the Edinburgh Botanic Garden.

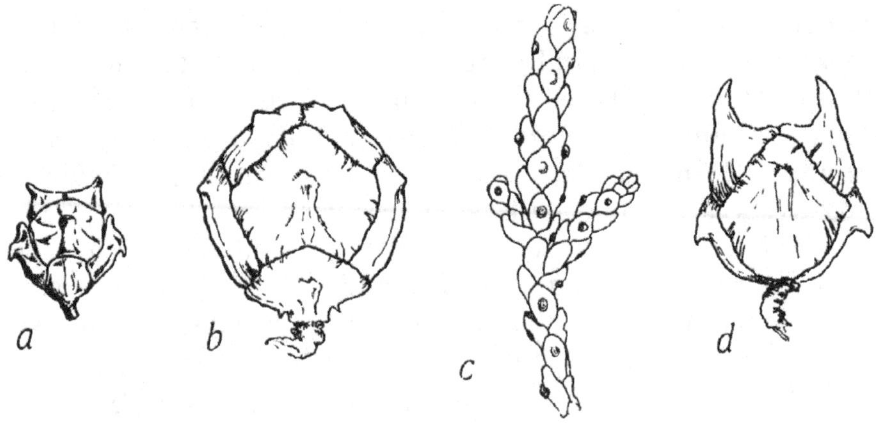

Fig 42 *a*, Modoc Cypress (*Cupressus macnabiana* var. *bakeri* Jepson), cone, nat size *b*, Sargent Cypress (*Cupressus sargentii* Jepson), cone, nat size *c*, McNab Cypress (*Cupressus macnabiana* Murr), branchlet showing glandular pits on back of leaves, 6 times nat size, *d*, cone of McNab Cypress, nat size

Modoc Cypress (var. bakeri Jepson; fig. 42a) is a shrub 6 to 10 feet high or becoming a small tree up to 25 feet. The bark is red-brown and the branchlets very slender. It grows on the lava beds of southeastern Siskiyou and southwestern Modoc at 4000 feet altitude where it occurs in association with scrub Yellow Pine, Knob-cone Pine and Sierra Juniper.

Piute Cypress (Cupressus nevadensis Abrams) is a small tree local on Piute Mountain, Kern County.

Sargent Cypress

Fig 42b

Sargent Cypress (Cupressus sargentii Jepson) is a shrub or small tree 5 to 12 feet high with grayish brown fibrous bark. It grows on mountain slopes and is distributed from Red Mountain (southern Mendocino County) to Mt. Tamalpais and southward to the Santa Cruz Mountains. Altitudinally it occurs chiefly between 2000 and 2300 feet. The Dutton Cypress (var. duttoni Jepson) occurs on Cedar Mountain in the Mt. Hamilton Range.

California Juniper

California Juniper (Juniperus californica Carr.) is a bushy shrub 2 to 15 feet high, or sometimes a tree up to 25 feet high. The bark is ashen gray or brown, the thin outer layers becoming very loose and shreddy. It inhabits arid or desert foothills and is most abundant on the western Mohave Desert, particularly on the desert slopes of the San Bernardino, Sierra Madre, Sierra Liebre and Tehachapi mountains. Thence it ranges westward to the San Rafael Mountains and northward it is scattered at intervals along the inner South Coast Range as far north as Mt. Diablo. In the North Coast Ranges it occurs from Lake County to the easterly foothills of the Yollo Bolly Mountains. Southward it is found along both slopes of the San Jacinto Range into Lower California, and extends northward in the Sierra Nevada to Kern River Valley as far as Kernville, recurring locally about Coulterville in Mariposa County.

Desert Juniper (var. utahensis Engelm.) is a small or stunted shrub 3 to 10 feet high or rarely a tree up to 20 feet high. It is very similar to the California Juniper but distinguishable by its more slender branches, usually glandless leaves and globose berries. It inhabits the desert ranges east of the Sierra Nevada (White, Inyo, Panamint, Providence and Grapevine mountains), and ranges through Nevada to the central Rocky Mountains and to northern Arizona. Its wood is hard, fine-grained and exceedingly durable.

In the sparingly wooded regions where this tree grows it is, like the California Juniper, an important resource to settlers for fence posts and fuel.

Sierra Juniper

Fig 43

Sierra Juniper (Juniperus occidentalis Hook.) is a subalpine tree 10 to 25 or sometimes 65 feet high with trunk 1 to 5 feet in diameter. It inhabits the Sierra Nevada, where it reaches its best development, and is a timber line tree at altitudes of 9000 to 10,500 at the south and 7000 to 9000 feet at the north. It also occurs at a few stations in the Yollo Bolly Range, San Bernardino Mountains and Panamint Range. Northward it extends through eastern Oregon and Washington to Idaho.

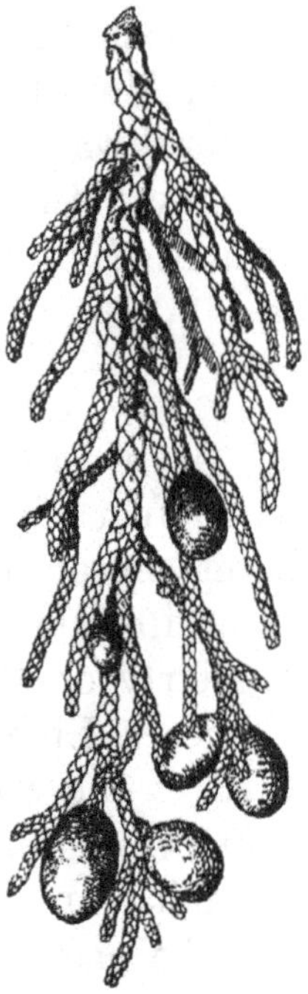

Fig 43 SIERRA JUNIPER (*Juniperus occidentalis* Hook.). Fruiting branchlet, nat. size.

The crown, which is a full and rather regular cone, is replaced by much broken or deformed tops wherever the trees grow in exposed situations. Since a favored habitat is high wind-swept granite plateaus or ridges, highly irregular or even prostrate crowns are a very characteristic feature of the subalpine region of the Sierra Nevada.

At lower altitudes in the same range Sierra Juniper is often a very conspicuous figure on granite tables, shelves and ledges of cañon walls or cliffs where it sometimes grows to great size, trunks 4 to 6 feet in diameter not being uncommon.

Western Yew

Fig. 44

Western Yew (Taxus brevifolia Nutt.) is a small tree 10 to 30 or rarely 50 feet high with an irregular crown, the branches of unequal length and standing at various angles, but tending to droop. The trunk is ½ to 2 feet in diameter with a thin red-brown smooth bark which is superficially deciduous in small thin shreds.

Taxus brevifolia inhabits deep cool shady cañons or stream-bottoms. The localities in California are comparatively few and rather widely separated. In the Coast Ranges it occurs in the Santa

Fig 44. WESTERN YEW (*Taxus brevifolia* Nutt) *a*, Fruiting branchlet, nat. size; *b*, longitudinal section of "berry," 1½ times nat size.

Cruz Mountains, on Mt. St. Helena, at various stations in Mendocino and Humboldt counties, on Marble Mountain and thence east to the Sacramento River cañon. In the Sierra Nevada it occurs from Butte County southward to Tulare County.

The wood of Western Yew is very fine and close-grained, very hard and heavy, flexible and remarkably durable. It is used by mechanics for tool handles and machine bearings and by the native tribes for their best bows. Yew logs, buried in the alluvial benches of the Eel River doubtless for several centuries, have been excavated and used by rural artisans for wedges and pulleys and by the settlers for mauls and gate-posts.

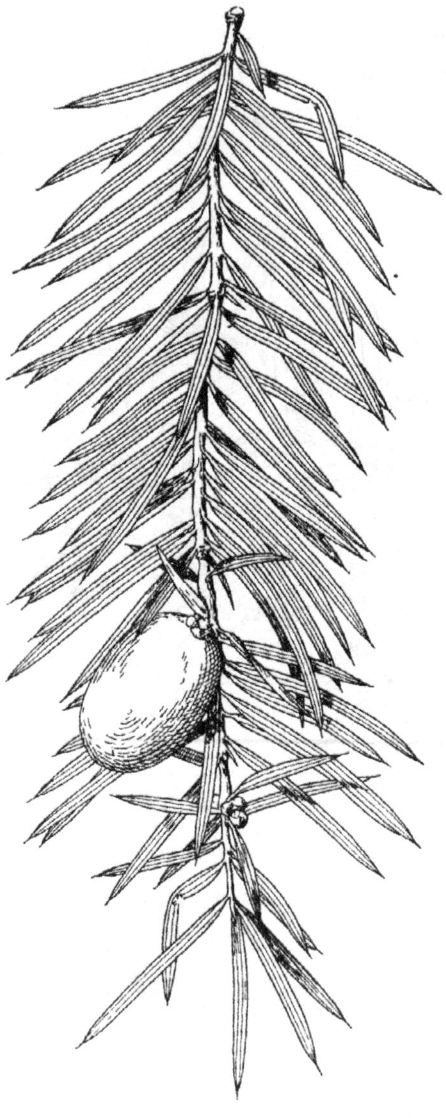

Fig 45 CALIFORNIA NUTMEG (*Torreya californica* Torr). Fruiting branch, ½ nat. size

California Nutmeg

Figs. 45, 46

California Nutmeg (Torreya californica Torr.) is a handsome tree 15 to 50 feet high with dark green foliage. The straight trunk is ½ to 3 feet in diameter with dark smoothish and thin bark. It is most readily recognized by its rigid bristle-pointed leaves spreading in two opposite rows and by its fleshy fruits which strikingly resemble a plum or olive.

Torreya californica inhabits cool shady cañons or sheltered slopes and is distributed in the Coast Ranges from the Santa Cruz Mountains north to Marin, Napa, southern Mendocino and Lake counties, and in the Sierra Nevada (west slope) from Tehama County to Tule River. While its range is fairly extensive the localities are comparatively few in number and the trees few in a locality. It is, for example, scattered along the Merced River, from El Portal to the lower Yosemite, one tree or bush about every one hundred yards on the average, but never in groves or even groups.

The wood is fine- and close-grained, elastic, rather heavy and very durable. It has been used by settlers for bridge timbers by

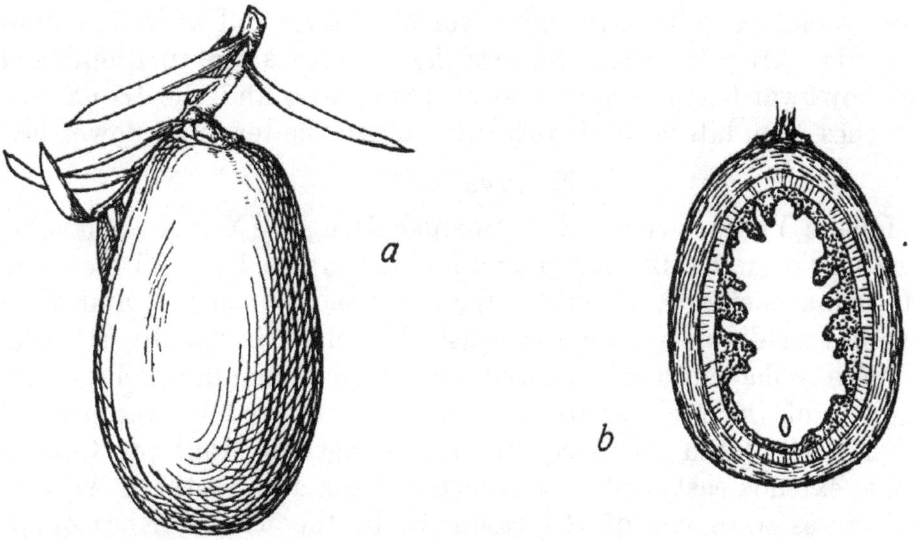

Fig 46 CALIFORNIA NUTMEG (*Torreya californica* Torr) *a*, The plum-like fruit; *b*, longitudinal section, showing the analogy of the fruit to that of the true Nutmeg of commerce. nat size.

virtue of its lasting quality, but being susceptible of a beautiful finish it would commend itself for many kinds of fine work did it occur in commercial quantity.

The fruit in its internal structure is suggestive of the Nutmeg of commerce, Myristica fragrans of the tropics. This resemblance is however purely superficial, since neither in botanical nor in economic character are the fruits alike.

Joshua Tree

Joshua Tree or Tree Yucca (Yucca brevifolia Engelm.) is a peculiar tree commonly 20 to 30 feet high with an open crown of arm-like branches, the columnar trunk 8 to 15 feet high and 1 to 3 feet in diameter. It inhabits arid mesas and mountains and is widely distributed in the Mohave Desert, extends northward to the Kern River Valley and Coso Mountains, and thence ranges eastward through southern Nevada to southwestern Utah. The trees form in many places, notably on the Mohave Desert, scattered groves where they impart to the desert landscape a singularly weird appearance. The sense of reverential awe exerted by these desert yuccas upon the mind of the Mormon settlers in Utah found expression in their folk-name Joshua Tree.

The stem does not branch until after the first flowering and is densely clothed with stiff spiny serrate leaves, all of which point upwards. After the plant flowers the old leaves die, turn outwards and downwards, falling in one or two years, and the trunk then branches from lateral buds formed beneath the terminal flower bud.

Mohave Yucca

Mohave Yucca, often called Spanish Dagger (Yucca mohavensis Sarg.), is a cactus-like shrub or a low tree up to 10 or 15 feet high. The trunk is simple, or with a few very short branches, and about 6 inches in diameter; on the coast this plant is usually stemless.

Yucca mohavensis is scattered over deserts, mountain slopes and plateaus of the Mohave desert, thence southward to western San Diego County and northerly along the coast to Monterey County. It also extends eastward to southern Nevada and northern Arizona. The leaves were one of the resources of the native tribes of the desert for fibres, being manufactured into blankets and cords.

California Fan Palm

California Fan Palm (Washingtonia filifera Wendl.) is a columnar tree 20 to 75 feet high, the trunk unbranched, 1 to 3 feet in diameter at the enlarged base, covered with a scaly rind and sometimes clothed quite to the ground with a thatch of dead persistent recurved leaves. It grows along alkaline streams, rivulets or springs on the northwestern and western margins of the Colorado Desert (a one-time inland sea) and thence southward to Lower California. On account of its necessity for an abundant root-supply of water it is highly localized in its distribution, some stations carrying only a few trees, others as many as a thousand. The tree is however a favorite for avenues and parks, and is more widely planted in California than any other palm. The known localities in the desert are comparatively few in number.

To the tribes of the Colorado Desert this tree was of economic importance. They used the leaves to thatch their huts and one may still see Indian houses with roofs and sides of palm leaves at the place called Los Coyotas in Coyote Cañon, as well as at other native villages in the Santa Jacinto Range. The soft bases of the young leaves are eaten, while the three or four clusters of fruit borne by a tree, each weighing about ten pounds, are an important source of food supply.

White Alder

White Alder (Alnus rhombifolia Nutt.) is a tree 30 to 100 feet high with a thin or open crown, tall slender trunk 1 to 3½ feet in diameter, and smooth whitish or gray-brown bark. As an inhabitant of river banks or cañon streams it grows in the Sierra Nevada up to 2500 feet at the north and 6000 or 8000 feet at the south, follows the main rivers in the Sacramento and San Joaquin valleys, extends westward through the Coast Ranges to the edge of the narrow coast belt occupied by the Red Alder, and ranges southward into Southern California (San Gabriel, San Bernardino, San Jacinto and Cuyamaca mountains), and northward to the Cascades of Washington.

Alnus rhombifolia keeps to streams that are permanent and is to the traveler a more reliable sign of water than Sycamore or even Fremont Cottonwood, although of far less practical value than Cottonwood because it does not occur in as strictly desert country. The files of trees in mountain gorges are of distinct value as stream-cover, as well as enhancing the beauty of the cañons by their long slender white trunks and airy crowns.

The wood is light, brittle and coarse-grained; it warps and checks badly when sawn so that it is seldom milled. Settlers make local use of the slender trunks for studs and rafters in barns and employ the larger ones for the construction of log houses.

Red Alder

Fig. 105

Red Alder (Alnus rubra Bong.) is a tree 25 to 90 feet high with large roundish or depressed crown, the trunk 40 to 60 feet tall and 1 to 2½ feet in diameter. The bark is usually very white-mottled, red on the inside.

Alnus rubra grows in the beds of cool cañon streams or on moist flats. It is confined to a narrow strip along the coast from the Santa Inez Mountains northward to Del Norte County, ranging far northward to Alaska. In Marin, Mendocino and Humboldt counties it forms pure groves of great beauty in bottom lands near the sea.

The wood is soft, brittle and coarse-grained. It is useful for fuel, and has been used for piles, bridge foundations and boats.

Water Birch

Water Birch, also called Black Birch and Red Birch (Betula occidentalis Hook.), is a slender tree or tall shrub 10 to 25 feet high with red-brown smooth bark and warty twigs. It grows sparingly along streams in cañons on the west slope of the Sierra Nevada from Bubbs Creek (8000 feet) northward to Mt. Shasta (2500 to 5000 feet), thence westward to the South Fork of the Salmon and to Humboldt County. It is most common with us along the water-courses on the eastern slope of the Sierra Nevada in the Owens Valley region where the poles are used for fencing.

Valley Oak

Fig 47

Valley Oak (Quercus lobata Neé), called Roble by the Spanish-Californians and Weeping Oak by the American settlers, is a graceful tree commonly 40 to 75 feet tall with a trunk 2 to 6 feet in diameter. The round-topped crown is often broader than high, its spreading branches finally ending in long slender cord-like branchlets which sometimes sweep the ground. The bark is dark brown, sometimes ashen gray, deeply checked into small rectangular or narrow or cuboid plates.

Quercus lobata is the most characteristic tree of valley levels in the Sacramento, San Joaquin and North Coast Range valleys and of many valleys in the South Coast Ranges. It is also found in mountain valleys in the Sierra Nevada foothills up to 2000 feet at the north and to 4000 feet at the south. Southward it ranges from Fort Tejon through the Santa Barbara mountains to the Ojai Valley; beyond this only in a feeble way as far as San Fernando Valley. In the coast country it sometimes approaches the ocean under the shelter of a protecting range but is never found in valleys facing the sea.

Quercus lobata is strongly marked by its deep tuberculate cup, long cartridge-like nuts, pinnately lobed leaves, cuboid-broken bark, and weeping sprays. These characters are very uniform throughout its range.

The root-system of this and other species of native oaks is host to the fungus Armillaria mellea (Vahl) Quel. When oak trees

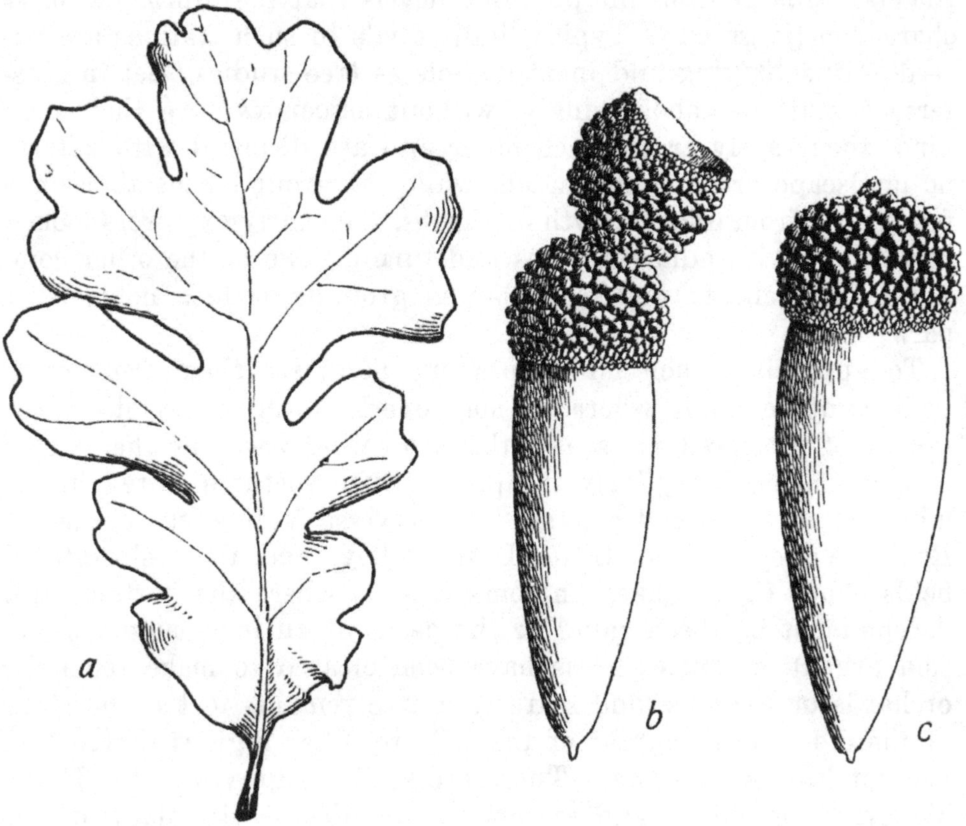

Fig 47. Valley Oak (*Quercus lobata* Neé). *a*, Typical leaf, *b*, *c*, acorns nat size

are cleared from land for orchard planting the roots of orchard trees become infested and death often results. When the roots of fruit trees come sufficiently near to the dead but infected oak roots in the soil the fungus grows over (by means of slender rhizomorphs) and attacks the living roots of the orchard tree. Until the roots of the oak are very completely decayed, the fungus does not die out completely and infection, says W. T. Horne, may result ten years after the clearing of oak trees from the land.

Quercus lobata attains its greatest development in the deep moist loam of alluvial or delta valleys, as on the Kaweah delta in the San Joaquin, about Marysville and Chico in the Sacramento, and in some of the North Coast Range valleys. In such places individual trees not infrequently attain diameters of 8 to 10 feet and extreme heights of 100 to 150 feet. Although growing in smaller form on the clay hills, it is on the plain-like levels that they form the most characteristic growth. Typical individuals in such stations are exceedingly satisfying and inexhaustible as tree studies. Set in clusters or scattered about singly, without associates save their own kind, they rarely crowd each other and are disposed with a taste no landscape artist could match, while the ground beneath is perfectly free from undergrowth or shrubs. The early explorers looked upon them with admiring eyes and turning one to the other compared the scene to well-kept planted grounds or to a nobleman's park.

To appreciate such landscapes one must see them from some little vantage point where he may overlook the valley floor, the groves of scattered trees and the projecting bases of the purple hills indenting irregularly the plain. Such pictures stored in the mind, recall the broad expanse of Berryessa Valley, the circles of Round Valley and of Little Lake Valley, and the well-watered fields about Clear Lake. In some regions where the horticultural development has been rapid or the needs of an increasing population urgent, extensive areas have been cleared to make room for orchards or gardens, and scarcely a tree remains to tell the story of the old time monarchs of the soil; in other regions destruction has not been so complete. There are still fine groves in the Ukiah Valley, Napa Valley and in certain localities on the plains of the Sacramento. The valleys about the base of Mt. Diablo still hold semi-primitive clusters, as do other of the inner Coast Range valleys further removed from centers of population. Old Fort Tejon, at the head of the San Joaquin, is set in an interesting assemblage of large-sized trees.

Valley Oak is of little economic importance except as a shade tree in farming fields or as to the use made of it for fuel. Its wood is white, hard and brittle, being the least esteemed hard wood of California. Most evidence goes to show that it rots quickly in contact with soil, although posts made from the butt cut are said in some cases to last thirty years. While extensively used for firewood it is rarely employed for any other purpose. So frequently an inhabitant of delta lands it is called "Water Oak," "Bottom Oak" or "Swamp Oak," while the folk name "Mush Oak" carries with it a species of contempt and tells the story of its failure to meet the requirements of a tough strong wood in a land where good oak is scarce and dear.

Appreciation of the Valley Oak by the traveler must, therefore, rest almost wholly upon sentimental grounds. The leaves do not fall until late December, but defoliation changes little the aspect of a tree which makes so slight concession to the seasons. On the rich valley levels these trees are never mistaken. Whether it be in the still summer days with jays, woodpeckers and crows noisy in their tops, in the full of the hunter's moon with their bulks rising darkly out of the white stubble fields and the delicate fragrance of the foliage filling the air, or in the blackness of winter night when the north wind is shouting across the plain and their massive branches are traced against the bright glow of tule fires in the river bottoms —at all seasons their charm is in their tall broad crowns, their story on story of tortuous branches, their graceful drooping sprays and the distinct individuality which resides in every tree. They are, as the wise first-comers well knew, the sign of the richest soil. They tell the mettle of the land and they give the land a fine distinction

Oregon Oak

Figs. 48, 49

Oregon Oak (Quercus garryana Dougl.) is a tree 25 to 55 feet high with a rounded crown and trunk 1½ to 5 feet in diameter. The trunk bark is white, thin (½ inch thick), smoothish but superficially fissured into longitudinal bands which are transversely checked into small squarish scales 1 inch or less broad. It is best known by its broad spreading crowns—often broader than high, its white squarish-scaly trunk bark, mossy main arms and glossy leaves. It inhabits mountain slopes, ridges and cañons and is distributed from the Santa Cruz Mountains to Mt. Tamalpais (north slopes), and northward in the Coast Ranges through Oregon and Washington to British Columbia. In Sonoma, Mendocino and Humboldt counties it is abundant in the "Bald Hills" country inside the Redwood Belt where in company with Douglas Fir, Tan

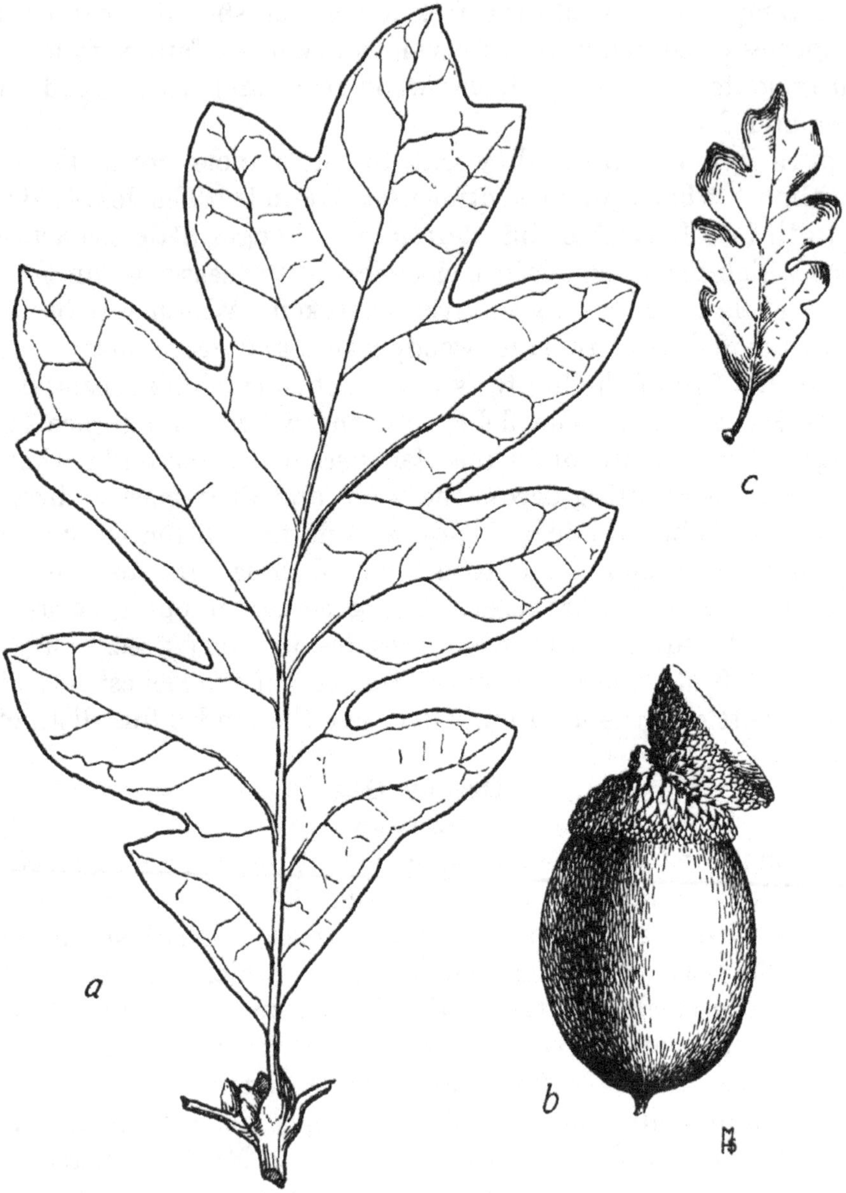

Fig 48. Oregon Oak (*Quercus garryana* Dougl). *a,* Leaf, the type with acutish lobes; *b,* acorn. *c,* Brewer Oak (var. *breweri* Jepson), typical leaf. nat. size.

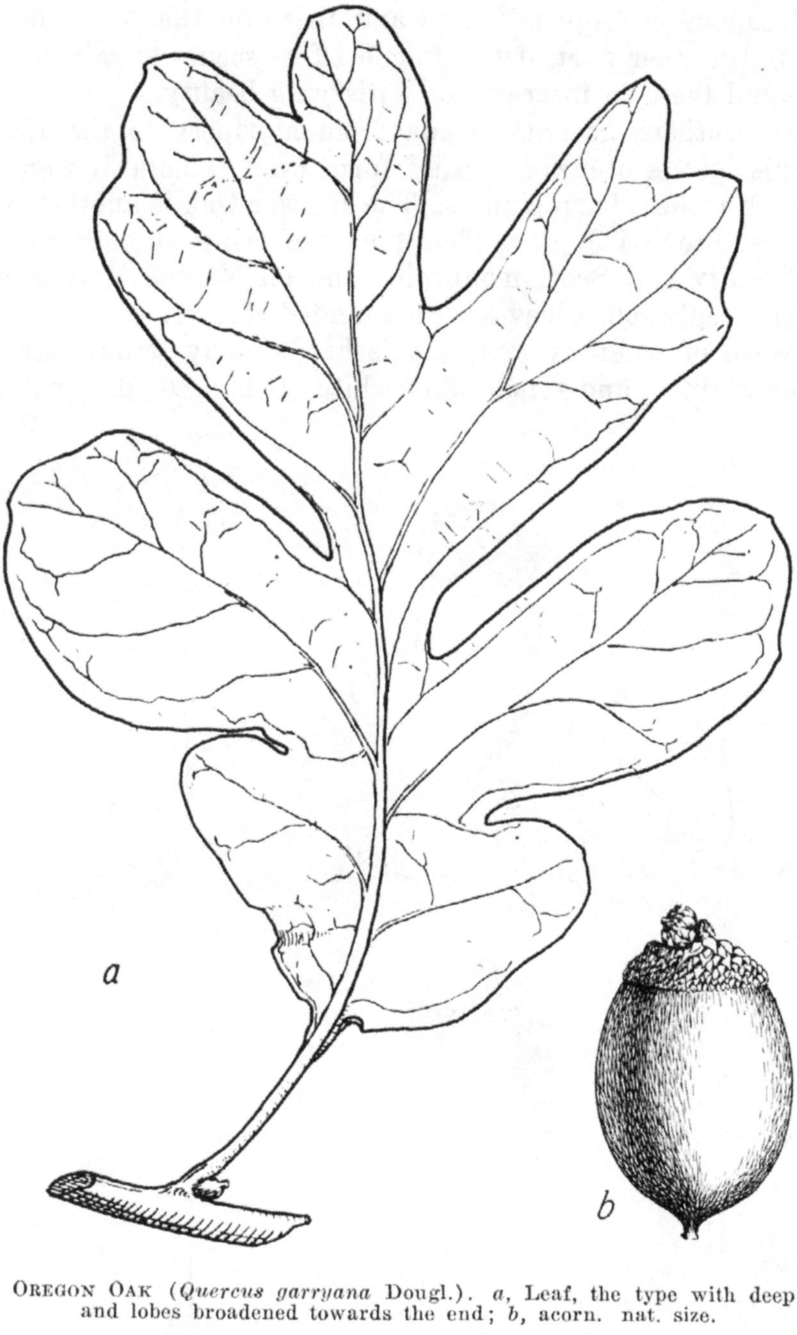

Fig. 49. OREGON OAK (*Quercus garryana* Dougl.). *a*, Leaf, the type with deep sinuses and lobes broadened towards the end; *b*, acorn. nat. size.

Oak and Madroño it forms extensive groves or small irregular clusters with grassy deer-parks between, free glades at the head of wooded cañons or "opens" here and there on the slopes or very summits. No other part of California offers scenes of mixed woods which equal these in interest and satisfying beauty.

In the southern Sierra Nevada from Mariposa to the Kaweah Basin this species occurs in dwarf form (var. semota Jepson), the leaves with rather sharp sinuses. The Brewer Oak is another dwarf form (var. breweri Jepson, Fig. 48c) occurring at high altitudes in the Trinity and Scott mountains and on Marble Mountain. It has small deeply lobed leaves with rounded sinuses.

The wood of Quercus garryana is hard, fairly strong, straight- and close-grained, and remarkably white. It is used for fence posts

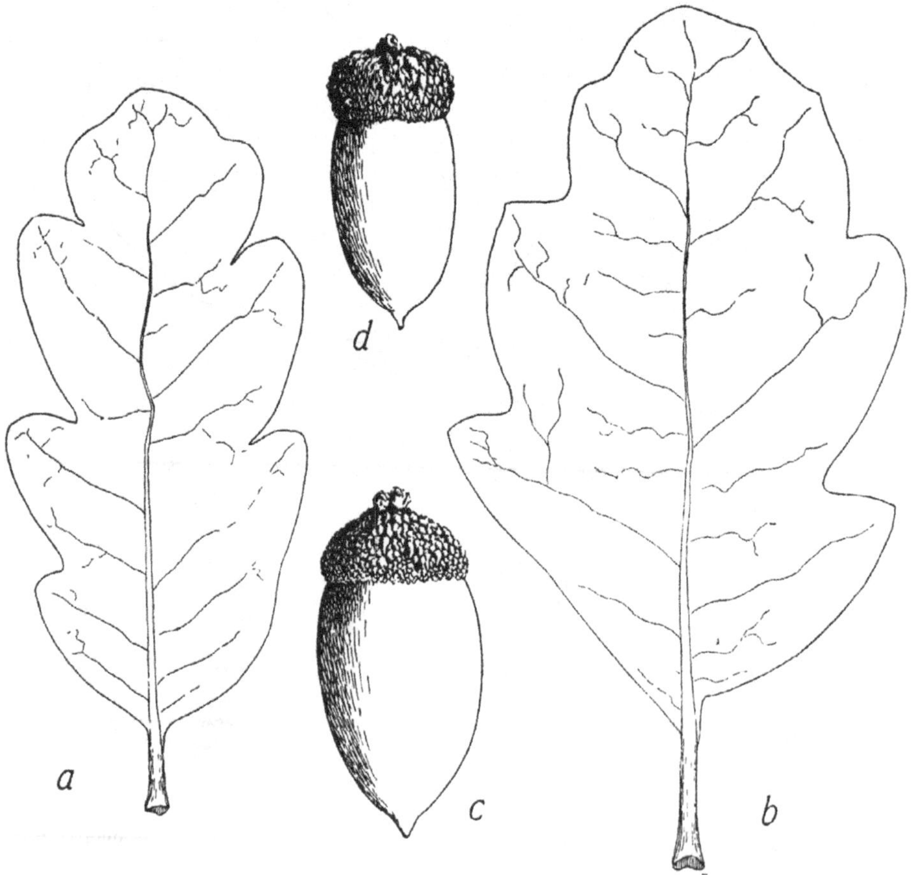

Fig 50. BLUE OAK (*Quercus douglasii* H. & A.). *a, b,* Leaves: *c. d.* acorns. nat. size.

(therefore called Post Oak) and in Oregon for furniture and interior finish. It is, next to Maul Oak, the most valuable of West American oaks.

Blue Oak

Figs 50, 51

Blue Oak (Quercus douglasii H. & A.), also called Mountain White Oak or Rock Oak, is a tree 20 to 60 feet high with a rounded crown and trunk 1 to 2 or sometimes 4 feet in diameter. The bark on the main trunk is white and shallowly checked into small thin scales; it is only slightly roughened but with the characteristic roughness extending up the limbs well onto the branches. It is most easily recognized by its white trunks and blue foliage.

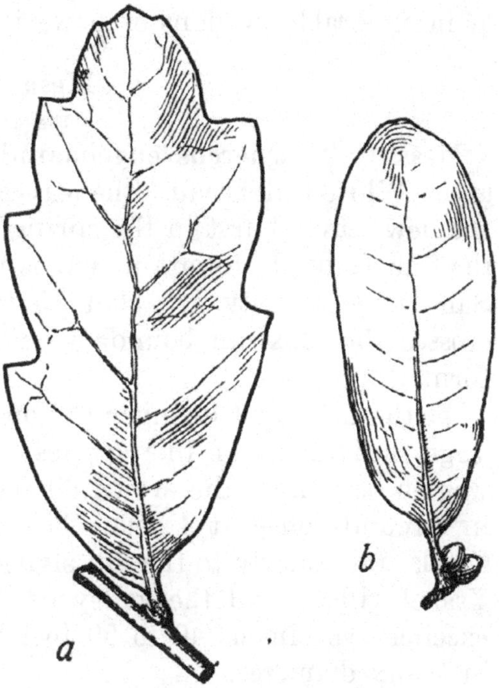

Fig 51 BLUE OAK (*Quercus douglasii* H. & A.), types of leaves *a*, Toothed leaf, *b*, entire leaf nat. size.

Quercus douglasii inhabits rocky or clay hills and is widely distributed through the foothill country around the Great Valley, north to the upper Sacramento and south to Fort Tejon. South of Tejon it occurs locally in the Sierra Liebre and in San Fernando Valley. In the Sierra Nevada foothills it is the most characteristic oak between 300 and 1500 feet at the north and 500 to 2500 or rarely 4000 feet at the south, forming groves of much-scattered trees and usually growing by itself, although often associated with Digger Pine and Interior Live Oak. In the Coast Ranges it is common in the inner and middle ranges, extending west to but not entering the Redwood belt and also avoiding the North Coast Range mountains of higher altitudes.

Its wood is close-grained, hard and brittle. Sometimes heartwood is so dense and hard that it will turn the edge of an axe, whence the settlers' name, Iron Oak. It is extensively used for firewood or occasionally for tool-handles. The acorns provide irregular crops as feed for hogs.

Not in itself an attractive tree the Blue Oak by reason of its form, color, and habit plays a strong and natural part in the scenery of the yellow-brown foothills. Always scattered about singly or in open groves, the trees are well associated in memory with bleached grass, glaring sunlight and dusty trails, although for a few brief days at the end of the rainy season the white trunks rise everywhere from a many-colored cloth woven from the slender threads of innumerable millions of flowering annuals.

Mesa Oak

Fig 109

Mesa Oak (Quercus engelmannii Greene) is often called Evergreen White Oak because the leaves persist through the winter until the new leaves burst in the spring. It occurs on the south slope of the San Gabriel Mountains but is most common on the low hills of San Diego County at about 15 to 30 miles from the ocean. It crosses the Mexican boundary and inhabits adjacent Lower California.

At the head of Clevinger Cañon not far from Ramona, are some large individuals of this species. The writer measured one which was 45 feet high, the crown 60 feet broad and the trunk 9½ feet in circumference at 4 feet. This species also grows along Witch Creek and easterly to the hill slopes overlooking the valley of Santa Ysabel village and the valley of Warner Ranch. Here are found excellent specimens 40 to 50 feet high and 1 to 2 and even 3 feet in trunk diameter.

Maul Oak

Fig 52

Maul Oak or Cañon Oak (Quercus chrysolepis Liebm.) is a tree 15 to 60 or even 110 feet high with roundish, often spreading crown and trunk 1 to 5 feet in diameter. On high ridges or exposed mountain summits it is often reduced to a mere shrub a few feet high. Whitish bark, small entire and toothed leaves on same twig, old leaves lead-color beneath and young leaves yellow-powdery beneath, acorns with turban-like cups—these are the most striking characteristics of Maul Oak and by means of which it may most readily be recognized.

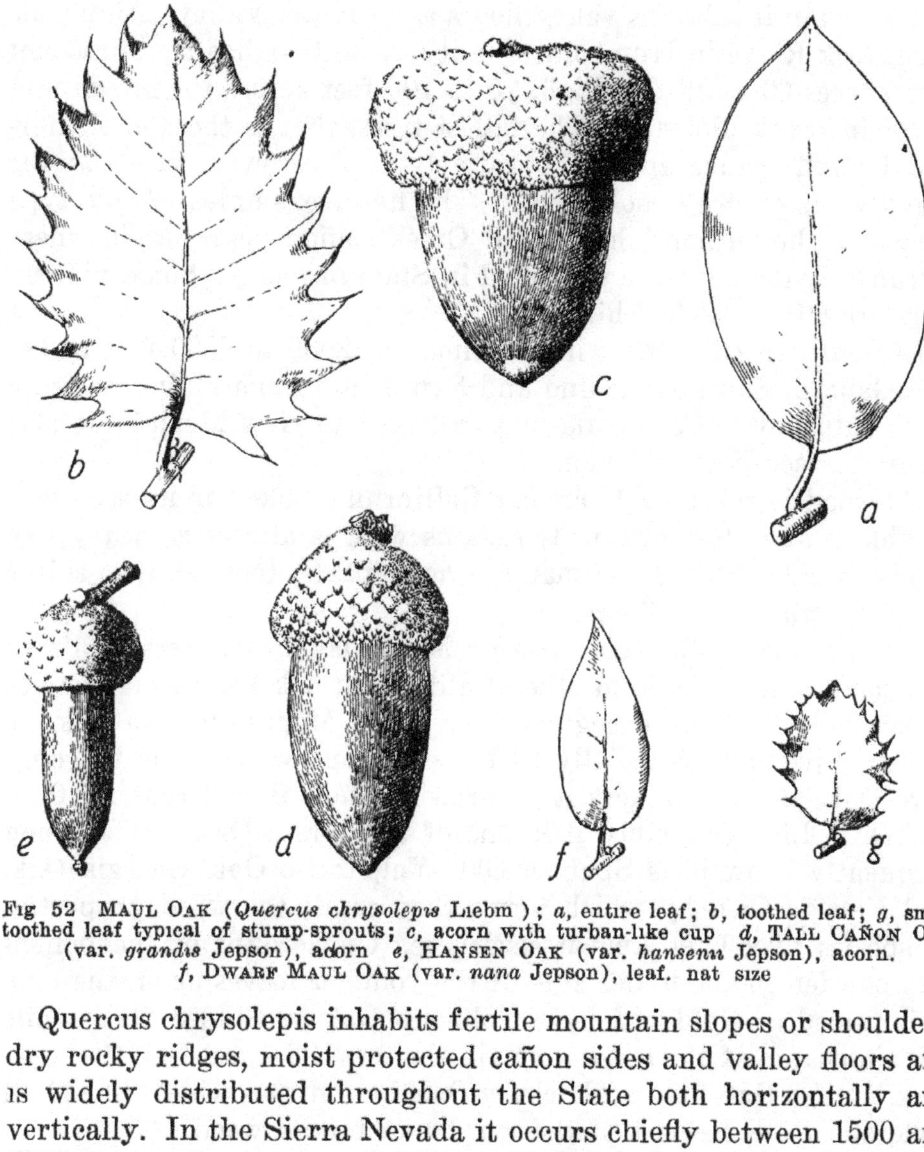

Fig 52 MAUL OAK (*Quercus chrysolepis* Liebm); *a*, entire leaf; *b*, toothed leaf; *g*, small toothed leaf typical of stump-sprouts; *c*, acorn with turban-like cup *d*, TALL CAÑON OAK (var. *grandis* Jepson), acorn *e*, HANSEN OAK (var. *hansenii* Jepson), acorn. *f*, DWARF MAUL OAK (var. *nana* Jepson), leaf. nat size

Quercus chrysolepis inhabits fertile mountain slopes or shoulders, dry rocky ridges, moist protected cañon sides and valley floors and is widely distributed throughout the State both horizontally and vertically. In the Sierra Nevada it occurs chiefly between 1500 and 5000 feet at the north and 3000 and 8000 feet at the south, mainly on cañon slopes where it is a low spreading tree often forming the main part of the cover. In Hetch-Hetchy, Yosemite, Kings and Kern, round balls of Maul Oak are a feature of the rocky walls and talus. In the Coast Ranges it is confined for the most part to the higher ranges and reaches its greatest development in Mendocino and Humboldt counties where large-sized broad-crowned trees dignify little shoulders on mountain slopes in the Bald Hills country, or again it inhabits valley floors as in Hupa Valley. Along the Blue Rock Ridge in Humboldt County the author has measured not a few trees 60 to 80 feet high, 70 to 100 feet across the crown and 3 feet in trunk diameter. The tallest trees, that is those characterized by tall trunks and comparatively small crowns, grow on the sharp walls of deep cool cañons. The finest examples of this type known to the author inhabit Mill Creek Cañon near Ukiah, where fortunately the trees are protected in State property. Some of these trees are 80 to 110 feet high.

In Southern California it is common at 2000 to 6000 feet in the San Gabriel, San Bernardino and San Jacinto mountains, ranging south into Lower California and eastward to New Mexico. It also occurs in southern Oregon.

Its wood is remarkable among Californian oaks for its strength, toughness and close grain. It seasons well, is almost as heavy dry as green, and is used for mauls, wagon parts, tool-handles, ship's knees, furniture and floors.

On account of the wide use of Maul Oak by settlers, millmen and mountain packers no other Californian oak has so many folk-names in use. Being evergreen it is called Mountain Live Oak, or merely Live Oak, especially in those regions where it is the only Live Oak: in other places it is termed Cañon Oak, Drooping Oak, or White Live Oak since it is one of the White Oaks. Woodsmen frequently know it as Spanish Oak, Valparaiso Oak, Georgia Oak, and Florida Oak, by which terms they recall the wood properties of species they have known elsewhere. On account of the pollen-like powder on the under side of the younger leaves or on the cup, it is famed as Gold-leaf Oak, Gold-cup Oak or Golden Oak, while certain shapes of the leaves explain the term "Laurel Oak." "Iron Oak," "Pin Oak," and "Hickory Oak" are names which, like Maul Oak, speak the respect of the ranchman for its wood.

Maul Oak is exceedingly variable in all its characteristics. The crown may be very tall and broad in the open, narrow with tall trunks in cañons, or with very low broad crowns and exceedingly short trunks on arid slopes. The leaves are very variable, especially as to the margins. Variability is also a marked characteristic of both acorn cups and nuts. Coloration is a convenient means of recognizing the tree as indicated above. Some of the more striking variations in leaves and acorns are shown in Fig. 52.

Maul Oak is a remarkable host for insect galls. While oak-galls, or "oak-balls," are common on our species of oaks, large-sized ones being characteristic of Quercus lobata and Quercus douglasii, it is on Quercus chrysolepis that the largest number of gall species are found. Up to the year 1918 L. H. Weld had determined some 54 species parasitic on the Maul Oak.

Island Oak

Island Oak (Quercus tomentella Engelm.) is a tree 25 to 55 feet high with roundish crown and trunk 1 to 2 feet in diameter. The bark is gray-brown, smoothish, with irregular flattish ridges separated by longitudinal fissures. It is a strictly insular species, first discovered on Guadalupe Island and since found on Santa Cruz, Santa Rosa, San Clemente and Santa Catalina islands. On Santa Catalina Island we saw only trees that had been derived from crown-sprouts—slender poles 40 to 55 feet high in clusters of 3 to 10 with diameters of 1½ to 2 feet. Quercus tomentella is the rarest in individuals of all West American oaks. It is scarcely more than a subspecies of Quercus chrysolepis but is remarkable for its tomentose leaves and the size of its acorn cups.

Coast Live Oak

Figs 53, 54

Coast Live Oak (Quercus agrifolia Neé), called simply Live Oak, is a low broad-headed tree commonly 20 to 40 but sometimes 70 feet high. The trunk is 1 to 4 feet in diameter, usually short and parting into wide-spreading limbs which often touch or trail along the ground. The trunk bark is smooth and beech-like, sometimes irregularly fissured or with an occasional very deep fissure.

Quercus agrifolia grows on rich valley floors, rocky hills, fertile slopes or benches in the hills, or on dry mesas. It is distributed in the North Coast Ranges from northern Sonoma County to Marin County and to Suisun Valley, and throughout the South Coast Ranges, where it is very abundant and widely scattered, to Southern and Lower California.

It is a hardy tree and is often the only tree which inhabits outlying or wind-swept stations in the South Coast Range country. It accommodates itself to such places as wind-gaps in the hills by developing its crown to leeward or on the exposed ridges by developing horizontally over the ground. As an indicator of the prevalence

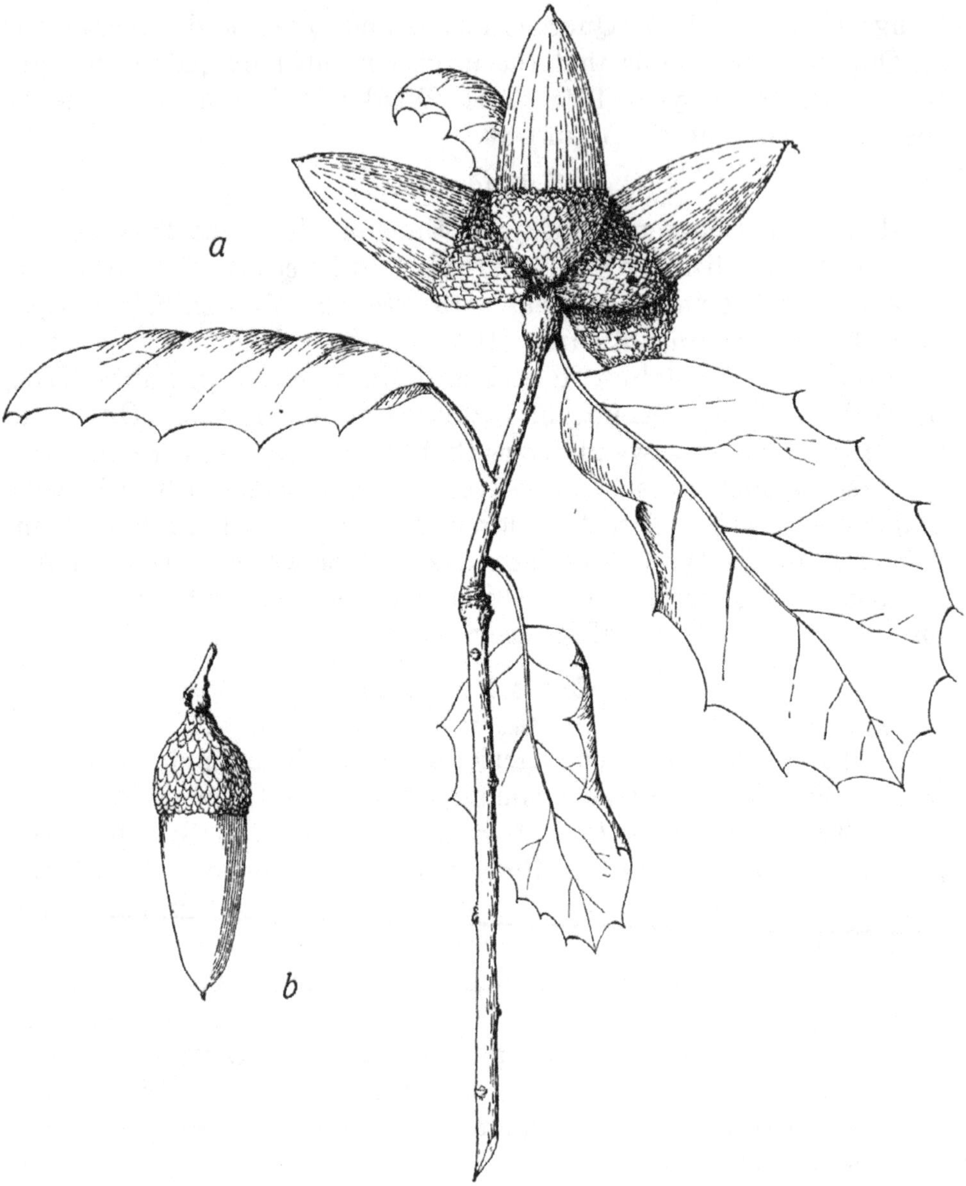

Fig 53 Coast Live Oak (*Quercus agrifolia* Neé). *a*, Acorn-bearing branch; *b*, the most common type of acorn nat size

and velocity of the trade winds in a particular locality no other tree gives such sure testimony. Some fine examples of crowns developed wholly to one side of the trunk may be seen between Carmel and Pescadero Point on the Seventeen Mile Drive.

Coast Live Oak is almost universally known among the people simply as Live Oak, or rarely as Field Oak. It is sometimes called Holly-leaf Oak or Holly Oak, a rather happy common name but lacking the force of usage. The specific name, agrifolia, is simply one of the forms of the old Latin name of the holly.

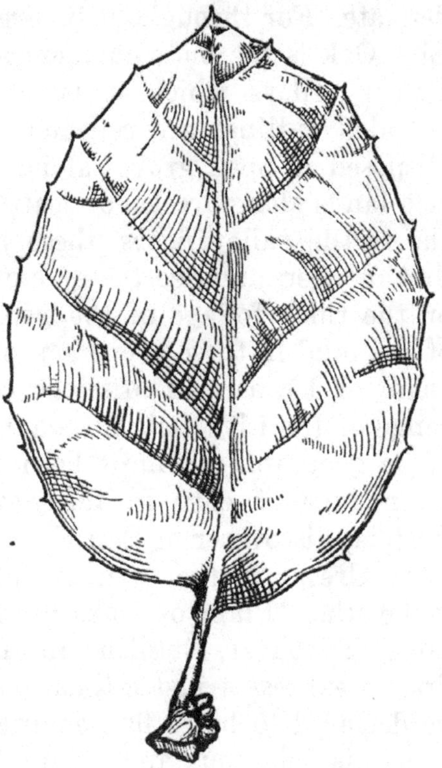

Fig 54 Coast Live Oak, usual convex type of leaf nat size

This species was first collected by the Malaspina Expedition in 1791, the first and last scientific expedition sent by Spain to her remote province of Alta California. The botanists of the expedition were Thaddeus Haenke and Luis Neé. The ship anchored for some time at the port of Monterey, where Haenke botanized, Neé having stopped in Mexico until the return of the expedition from the north. Specimens of two oaks were, however, gathered by ship's officers and given to Neé, who published at Madrid a description of them in the Anales de Ciencias Naturales. In his account of Quercus agrifolia he says that he cannot give the height of the tree of which he has seen only branches gathered by Don Joseph Robredo and Don Manuel Esquerra, who also collected the Valley Oak. These were the first California trees to be made known in Europe and of the broad-leaved forms these two species are in many ways the most interesting.

The wood of Quercus agrifolia is hard, heavy and moderately strong, is extensively converted into firewood and charcoal, and has been used for ship's knees and wagon parts. In a continued succession of drought years branches for browsing are cut from the trees to save range cattle from starvation.

This species, of inferior timber value, is undoubtedly of most economic importance to the community in an indirect way, that is in its relation to the heightened landscape effect of valleys and hills in a region which would in the main, save for it, be treeless and desolate. For throughout its area, except in the extreme north, this Live Oak is the most common and characteristic tree of the Coast Range valleys which it beautifies with low broad heads whose rounded outlines are repeated in the soft curves of the foothills. Disposed in open groves along the bases of low hills, fringing the rich lands along creeks, or scattered by hundreds or thousands over the fertile valley floors, these trees were of signal interest to the first pioneers; and so the eyes of the early Spanish explorers dwelt on the thick foliage of the swelling crowns and read the fertility of the land in these evergreen oaks which they called Encina. The chain of Franciscan Missions corresponded closely to the general range of the Live Oak, although uniformly well within the margin of its geographical limits both eastward and northward. The vast assemblage of oaks in the Santa Clara Valley met the eye of Portola, discoverer of San Francisco Bay, in 1769, and a few years later, Crespi, in the narrative of the expedition of 1772, called the valley the "Plain of Oaks of the Port of San Francisco." Then came Vancouver, Englishman and discoverer. Although he was the first to express a just estimate of the Bay of San Francisco, which he declared to be as fine as any port in the world, nevertheless it is in his felicitous and appreciative description of the groves of oaks, the fertile soil (of which they were a sign), and the equable climate that one reads between his lines of 1792 the prophecy of California's later empire.

Interior Live Oak

Figs 55, 56

The Interior Live Oak (Quercus wislizenii A. DC.), which in general appearance closely resembles the Coast Live Oak, is a tree 30 to 75 feet high with full rounded crown and trunk 1 to 3 feet in diameter. It inhabits rich valley floors, clay hills, rocky slopes or ridges, and is widely distributed through the Sacramento and San Joaquin valleys, especially on the east side, extending into the Sierra Nevada foothills to altitudes of 2000 feet at the north and

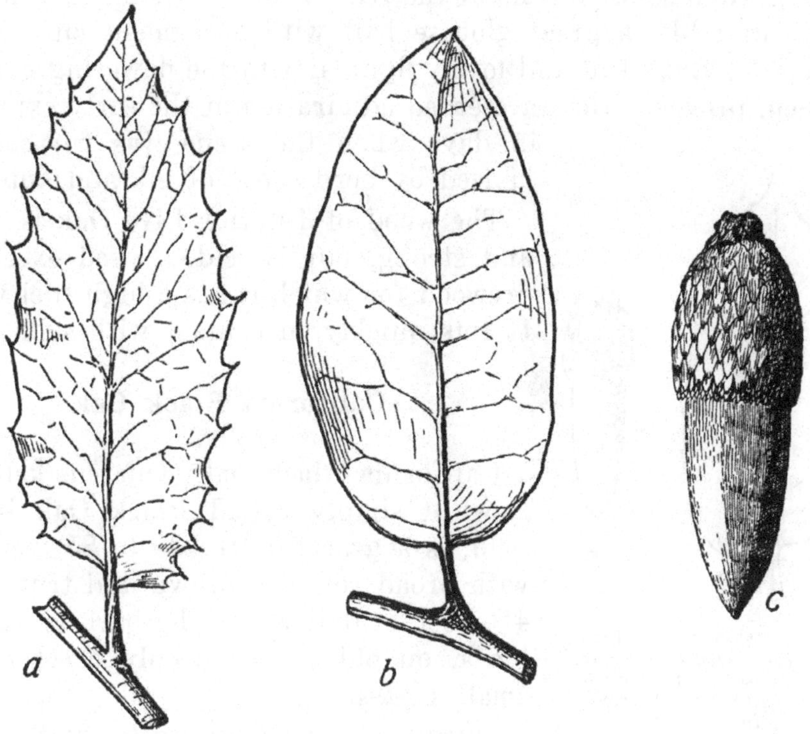

Fig 55 Interior Live Oak (*Quercus wislizenii* A DC). *a*, Typical spiny leaf; *b*, typical entire leaf; *c*, acorn. nat size.

4000 feet at the south. It also occurs in the inner North Coast Ranges, extending west to the Ukiah Valley.

The trunk bark is dark, very smooth or sometimes roughly fissured. A characteristic feature is the density of the periphery of the crown due to the abundant twigs and foliage. Its leaves are flat, while those of the Coast Live Oak are a little cupped. Moreover, its acorns require two years to mature while those of the latter species ripen in the first autumn after flowering. These two species, though so much alike in botanical character, have mutually exclusive ranges, although the scrub form (var. frutescens Engelm.) of Quercus wislizenii often tops the summit of mountains overlooking valleys inhabited by Quercus agrifolia.

The areas of best development of the Interior Live Oak are the broad alluvial banks of rivers on the east side of the Great Valley, such as the American, Consumnes, Mokelumne, Stanislaus and Tuolumne. Typical trees in this valley region have numerous branches, erect in the top and spreading around the sides and down to the ground so as to conceal the trunk, the crown thus resting on the ground like a great globose ball with a segment cut off the lower side. Such full and lovely figures, with the flowering prairies between, provoked the unreserved admiration of the early explorers in days when the scene was further enhanced by herds both of elk and antelope.

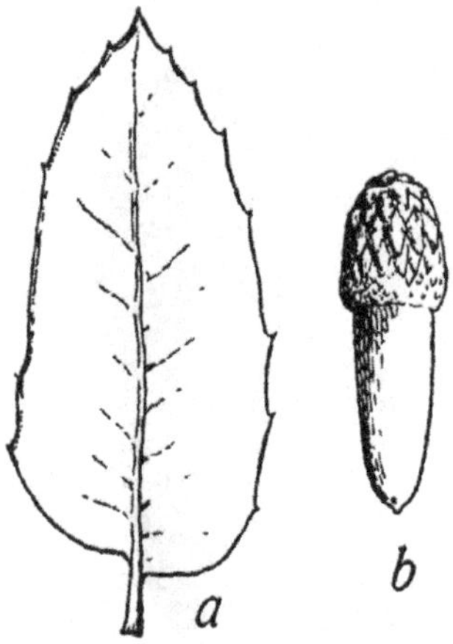

FIG 56 PENCIL-NUT LIVE OAK (*Q. wislizenii* var. *extima* Jepson) *a*, Leaf; *b*, acorn. nat size. Sierra foothills

The wood of Interior Live Oak is tough and strong, but is seldom used except as firewood for which it has a high fuel value. It rots quickly in contact with soil.

California Black Oak

FIG 57

California Black Oak (Quercus kelloggii Newb.), simply called Black Oak in the field, is a graceful tree 30 to 85 feet high with broad rounded crown and trunk 1 to 4½ feet in diameter. The bark is dark or black, on old trunks deeply checked into small plates.

Quercus kelloggii grows on high ridges, mountain slopes and in gravelly mountain valleys and is widely distributed through the Coast Ranges and Sierra Nevada, north to central Oregon, south to the high mountains of Southern California (San Gabriel, San Bernardino, San Jacinto, Palomar and Cuyamaca mountains). In the Sierra Nevada it is most common within or just below the lower margin of the Yellow Pine Belt at 1500 to 3000 feet at the north and 4000 to 6000 feet at the south. In the Coast Ranges it occurs chiefly between 200 and 4000 feet but its distribution is very scattered except in the higher North Coast Ranges of Mendocino and Humboldt inside the Redwood belt. It does not associate with Redwood and is not found near the sea nor usually on valley floors, except in such valleys as Santa Rosa, Napa or Ukiah where there are locally favorable spots of clay or gravelly soil. It is a most constant associate of Yellow Pine in the Coast Ranges and is commonly found with Oregon Oak, Tan Oak and Madroño. Black Oak, however, occurs in ranges where all of these species are absent, save rarely an isolated Madroño, as in the Vaca Mountains and on Mt. Diablo.

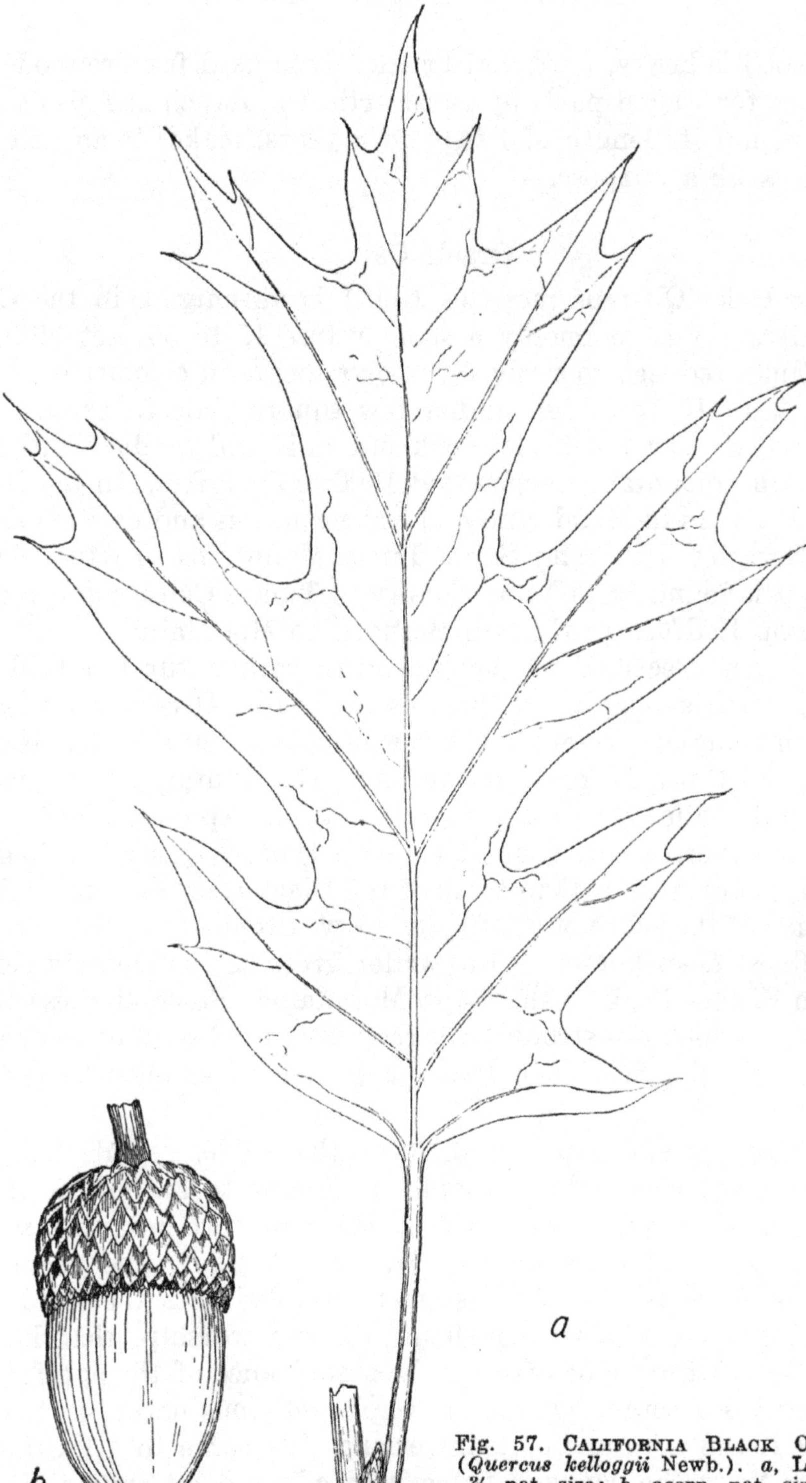

Fig. 57. CALIFORNIA BLACK OAK (*Quercus kelloggii* Newb.). *a*, Leaf ⅔ nat. size; *b*, acorn, nat. size.

The wood is heavy, hard and brittle. It is used for firewood and sometimes for wagon parts by rural artisans. Ranchmen also use it for posts, but its length of life, 4 to 8 years, makes it an inferior wood for such a purpose.

Oracle Oak

Oracle Oak (Quercus morehus Kell.) is an enigma in the California silva. It is commonly a slender tree 15 to 50 feet high, or is sometimes reduced to scrub form, or appears in colonies or dense thickets 7 to 10 feet high and a few square rods in extent. Its occurrence is very much localized, but it is widely distributed in the hills and mountains from about 1000 to 5000 feet. In the Coast Ranges it occurs in Mendocino and Lake counties and extends south to the Berkeley Hills and Santa Lucia Mountains. In the Sierra Nevada it is found from Yuba County to Tulare County and recurs near Camp Radford in the San Bernardino Mountains.

It is often described in the literature as a natural hybrid between Quercus kelloggii and Quercus wislizenii. It recalls the latter species in being evergreen, but in few other particulars. The leaves, which persist nearly or quite through the winter, are typically oblong and shallowly lobed with spiny teeth. Apparent leaf intergrades, however, occur, connecting with typical Quercus kelloggii. The acorns are smaller than those of the Black Oak, the cup slightly suggestive of the acorn cups of Quercus wislizenii.

The finest trees known to the writer grow on the easterly slopes of Twin Sisters Peak in the Napa Mountains. Some of these trees bear leaves which are strongly suggestive of the leaves of the genus Castanea; in the field such trees have been nicknamed "chestnut trees."

Dr. Kellogg, a devoted student of the Bible, called this tree Abram's Oak; which thus provides a clue as to the origin of his species name morehus. Moreh is a Hebrew word meaning soothsayer, says the Hebrew scholar, Dr. W. F. Badè, and occurs in several passages in the Old Testament coupled with the word oak, as naming some well-known shrine of tree worship, that is, the oak of the soothsayer or oracle. The significance of the application to our tree is obscure, but such a deeply religious nature as that of Kellogg found a means of transferring this name to a California species. As to the common name here adopted an explanation is also due. Many botanists have given more or less oracular opinions as to the hybrid or non-hybrid origin of Quercus morehus, but none has as yet subjected the problem to the test of successful experiment. It has therefore seemed apt to call our California tree the Oracle Oak.

Tan Oak

Figs. 58, 59

Tan Oak (Lithocarpus densiflora Rehd.) is a large tree, 50 to 150 feet high, in the open with broad crown rounded at summit, in dense forest with narrow pointed crown, the trunk 1 to 4 feet in diameter. It inhabits fertile mountain slopes and ridges and is distributed through the seaward Coast Ranges from Santa Barbara County north to Del Norte County and the Umpqua River in Oregon. In the North Coast Ranges it extends eastward to the Napa Range, Cobb Mountain, South Fork Mountain and New River. In the Sierra Nevada it occurs sparingly and in isolated localities from Butte County to Mariposa County.

The bark of Tan Oak furnishes the best tannage known for the production of heavy leathers, and extensive tanning plants are among the leading manufactories in California. The heavy leather tanneries consume about 20,000 cords of bark annually. The bark industry, an exploitation of the forests of Tan Oak, has been steadily growing. In taking off the bark in the peeling season, May to July, the men work in pairs. By a process which is called ringing, the bark is cut through to the wood in a circle around the trunk at the ground and again four feet above the ground, and then removed. This is the first "rim." The tree is then cut down. One man goes ahead "ringing" every four feet and the second man follows taking off the successive "rims." Bark less than three-eighths or one-half inch in thickness is commonly not peeled. Climbing a tree to get off extra rims is called "cooning." Young trees are peeled standing. In this case the bark is taken off as high as it can be reached with the axe, a method which is called "jay-hawking." Jayhawked trees live two seasons after removal of the bark. In the second season they commonly bear an exceptionally large crop of acorns and then die, a phenomenon aptly called by the woodsmen the "last kick" of the tree. Occasionally such trees may live on indefinitely.

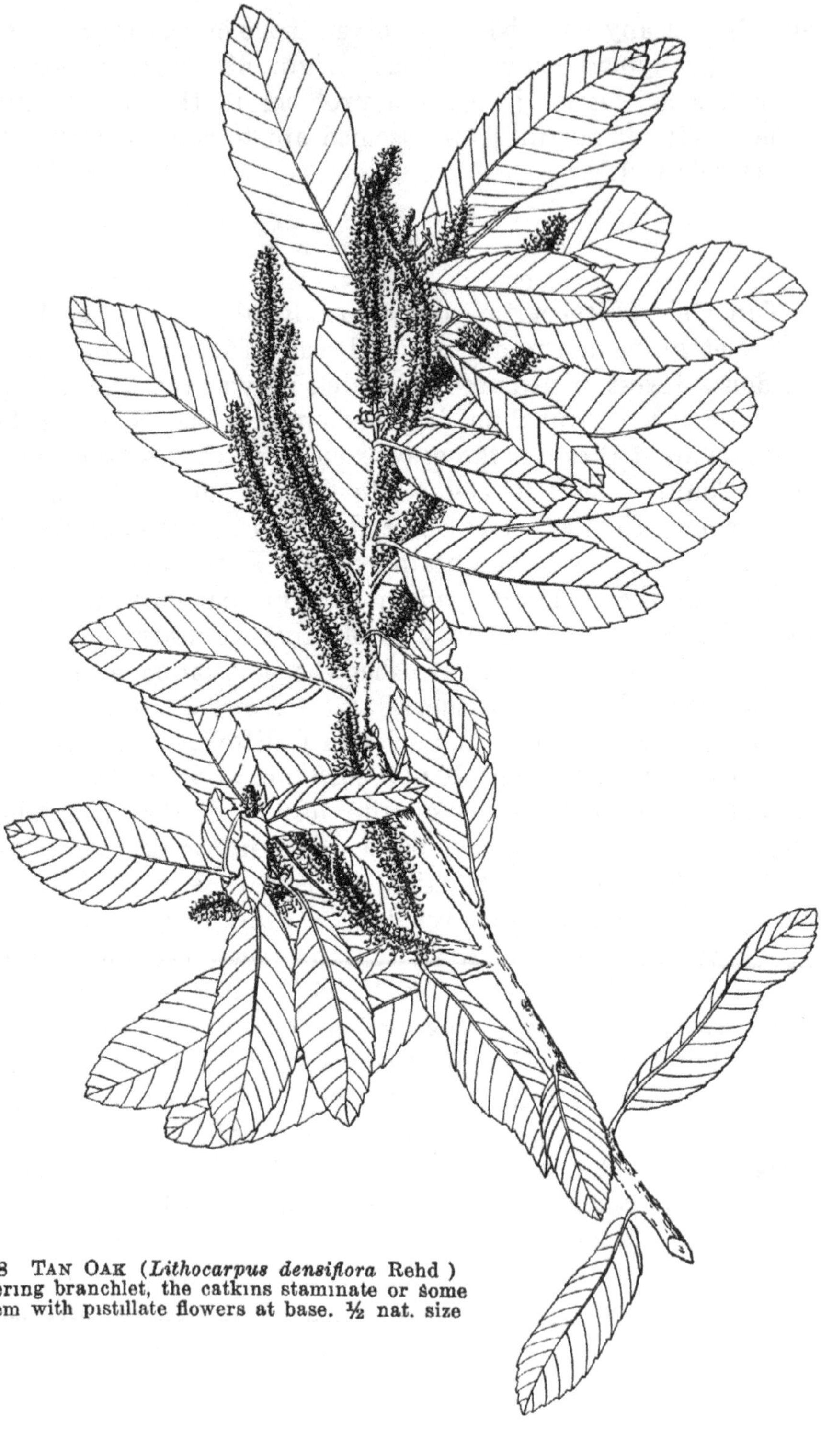

Fig 58 TAN OAK (*Lithocarpus densiflora* Rehd) Flowering branchlet, the catkins staminate or some of them with pistillate flowers at base. ½ nat. size

After the bark is stripped from the felled trees, about 100,000 trunks 10 to 100 feet long and ½ to 4 feet in diameter are left annually to rot on the ground, saving a small percentage, say 5 per cent., which is cut into firewood. It is believed that the wood has a high potential value, but commercial utilization is delayed because methods of handling it to the highest advantage have not as yet been discovered.

Tan Oak produces abundantly by crown sprouts. As many as fifteen to two hundred sprouts arise from the root-crown of a peeled tree. Eventually several shoots gain the ascendency and in fifteen or twenty years there is formed about the old stump a circle of slender pole-like trees. Sprouts may also be produced from the base of trees peeled standing. This fact fully explains the longevity, remarked above, which occasional trees with peeled trunks display, since the roots are thus furnished with assimilated food for growth without making a demand on the crown.

Reproduction is also provided for by seed. Acorns germinate readily in open spots in the forest where a great tree has fallen and permitted insolation. Seedlings are in fact a rarity because of too dense shade and the deep litter of undecayed twigs and leaves. Dry unprotected slopes are likewise discouraging to germination and especially trying to the vitality of the seedling in the rainless season.

Scrub Tan Oak, a low shrub 1 to 10 feet high, is a variety (var. echinoides Jepson) with thick entire leaves (1 to 2 inches long, the nerves inconspicuous), very bur-like cups and small roundish nuts. It occurs from near Mt. Shasta west to the Siskiyou Moun-

Fig. 59. TAN OAK acorns *a*, Pointed nut; *b*, *c*, the usual type of nut. nat size.

tains; also in the Sierra Nevada from Placer County to Mariposa County.

The genus Lithocarpus (Pasania) is represented by a single species in the New World, Lithocarpus densiflora, but some one hundred species occur in southeastern Asia and the Malay Archipelago. It is a genus about intermediate between the oaks and chestnuts. Lithocarpus has parallel lateral ribs to the leaf, erect staminate catkins with some pistillate flowers often borne near the base of the axis, and a somewhat bur-like cup. In all these features Lithocarpus resembles Castanea or chestnut, but the general habit and appearance of the tree is oak-like, while the fruit is essentially an acorn and not a chestnut bur. If, however, a nut be studied carefully from above, the observer will note that it is a little 3-angled, which again recalls the chestnuts.

Giant Chinquapin

Giant Chinquapin (Castanopsis chrysophylla A. DC.) is a forest tree 50 to 115 feet high with narrow pointed or in age rounded crown and tall trunks 2 to 6 feet in diameter. The bark is brown or dull gray on the surface, reddish inside, very fibrous, 1 to 3 inches thick and separated by deep longitudinal furrows into heavy rounded sparingly confluent ridges.

Castanopsis chrysophylla inhabits the deep soil of mountain ridges or slopes as an associate of the Redwood and is distributed from central Mendocino County, where it reaches its greatest development, northward to the Oregon Cascades. It is often called "Chestnut," and sometimes "Red Oak" or "Bur Oak," in the Mendocino woods.

Giant Chinquapin attains its largest form in the Mendocino woods, where it is a commercially rare tree, the giant individuals being scattered at wide intervals in the Redwood forest. Its wood is reddish-brown with white sapwood, close- and straight-grained and takes a glossy finish. It is rarely used for any fine purpose but is sometimes cut for fuel. The bark parts very readily from the wood and is used to adulterate Tan Oak bark, the fine logs being left to rot on the ground. This practice is all the more reprehensible in that the bark has no value for tanning purposes.

Golden Chinquapin (var. minor Benth., Fig. 116) is a shrub form with trough-like leaves very golden on the under surface. It ranges from Monterey and the Santa Cruz Mountains northward to Mendocino and Humboldt counties, growing chiefly on rocky slopes or ridges.

California Black Walnut

Fig. 60

California Black Walnut (Juglans hindsii Jepson) is a tree 50 to 75 feet high with massive crown and tall trunk 1 to 3 feet in diameter. On account of its different habit, different foliage, and larger nuts (1 to 1¾ inches in diameter) this tree is distinct from Juglans californica of Southern California.

As a native tree Juglans hindsii grows only in central California, being limited to a few isolated stations very much restricted in area. Such stations in every case represent old Indian village sites or camping grounds, and it seems likely that these trees may be survivals connected with the use of the nuts as food and their unconscious plantings by the native tribes. The known stations are as follows, the first two being represented by very large individuals: 1. Along Walnut Creek from near the east arm of Moraga Valley nearly to Pacheco, a few trees on Lafayette and San Ramon creeks; 2. Lower Sacramento River about Walnut Grove; 3. Napa Range, east slope near Wooden Valley; 4. Gordon Valley, one tree.

Juglans hindsii occurs too sparingly to furnish a supply of timber of commercial importance. Horticulturally the tree plays an important part in the walnut industry, since it is universally used

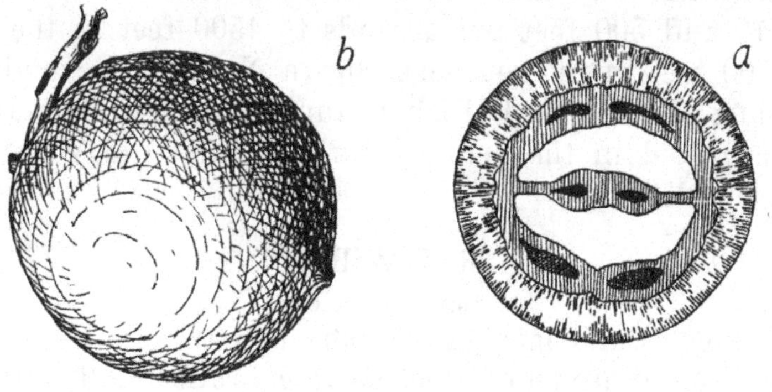

Fig 60. CALIFORNIA BLACK WALNUT. *b*, Nut; *a*, cross section nat size

as a stock graft for English Walnut, and is thus widely distributed over California as a cultivated tree. It is also grown as a windbreak, and very commonly as a roadside tree for which purpose it is admirably adapted.

Southern California Black Walnut

Southern California Black Walnut (Juglans californica Wats.) is a large many-stemmed shrub or tree 10 to 25 feet high, with roughish nearly black trunk bark. The foliage is aromatically pungent. It inhabits mountain slopes, stream beds or gravelly washes and is distributed from the Santa Maria watershed south to the Ojai Valley, Newhall, Santa Monica Mountains, eastward along the lower slopes of the San Gabriel and San Bernardino mountains and south to the Puente Hills and Brea Cañon in the Santa Ana Range (south limits).

Juglans californica is limited to Southern California. While individuals are often of large size, even of elephantine proportions, they are after all shrubs architecturally, since the trunk is very low-branching or is replaced by several stems from the base. Its nuts are small, being about 8 to 12 lines in diameter.

Yellow Willow

Figs 61c, d; 62b

Yellow Willow (Salix lasiandra Benth.) is a tree 20 to 45 feet high with a broad open crown of upright branches and brown roughly fissured trunk bark. The one-winter-old branchlets are yellowish, the winter buds short, blunt, and keeled on the back.

Salix lasiandra grows along the Sacramento and San Joaquin rivers and their tributaries, and fringes most Coast Range streams and creeks where the water flow is not intermittent; it occurs chiefly between 10 and 500 feet but ascends to 4500 feet in the northern and to 8500 feet in the southern Sierra Nevada. Beyond our borders it ranges north to British Columbia and Idaho. It is most easily recognized in the field by its glandular-warty petioles and long tapering leaves.

Red Willow

Figs 61a, b, 62c

Red Willow (Salix laevigata Bebb) is a tree 20 to 50 feet high with broad round crown of erect slender branches. The trunk bark is roughly fissured. The one-winter-old branchlets are reddish-brown with pointed ovate winter buds.

Salix laevigata grows along living streams or occasionally along summer-dry arroyos in regions of high winter precipitation and is

Fig 61 Red Willow (*Salix laevigata* Bebb); *a*, staminate catkin; *b*, pistillate catkin. Yellow Willow (*Salix lasiandra* Benth); *c*, staminate catkin; *d*, pistillate catkin. nat. size

distributed through the Coast Ranges, Great Valley and Sierra Nevada (especially the foothills) to Southern California. Beyond our borders it extends north to southern British Columbia. Altitudinally it ranges from near sea-level to about 4500 feet in the southern Sierra Nevada. It is commonly an associate of the Yellow Willow and has been variously called Bebb Willow, Smooth Willow and Spotted-leaf Willow.

Fig 62. *a*, BLACK WILLOW (*Salix nigra* Marsh.), leaf. *b*, YELLOW WILLOW (*Salix lasiandra* Benth), leaf. *c*, RED WILLOW (*Salix laevigata* Bebb), leaf. nat size.

Black Willow

Fig. 62a

Black Willow (Salix nigra Marsh.) is a tree 20 to 45 feet high with a roundish open crown of erect branches, the trunk with rough dark bark. It inhabits river banks in the Sacramento and San Joaquin valleys and follows the desert rivers through southeastern California and across southern Arizona to New Mexico and thence eastward to Texas and the Mississippi Valley, ranging as far north as Lake Superior and New Brunswick. It has a more extensive range than any other tree in the United States except the Aspen and is with the Aspen one of the two California trees in common with the silva of Eastern North America.

Arroyo Willow

Figs 63a, b, e; 64

Arroyo Willow (Salix lasiolepis Benth.), sometimes called White Willow, is a shrub or tree 8 to 20 or rarely 35 feet high with an irregular crown of spreading branches. The trunk is 3 to 9 inches in diameter, the bark smooth or on old trunks shallowly seamed.

Salix lasiolepis is distributed throughout the Coast Ranges, Great Valley and Sierra Nevada foothills, thence southward into Southern and Lower California. It is also reported from Arizona.

Growing along living streams in the valleys, Arroyo Willow also follows intermittent water-courses into the dry hills where it

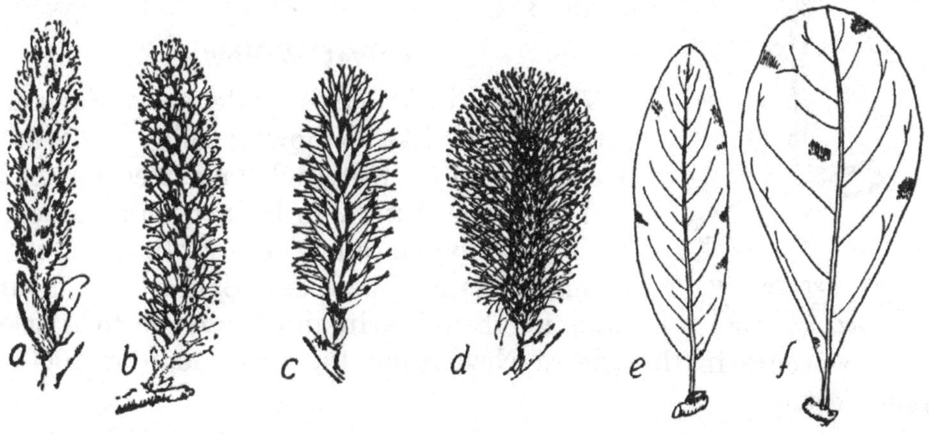

Fig. 63. ARROYO WILLOW (*Salix lasiolepis* Benth.); *a*, pistillate catkin; *b*, staminate catkin; *e*, leaf. NUTTALL WILLOW (*Salix scouleriana* var. *flavescens* Schn.); *c*, pistillate catkin; *d*, staminate catkin; *f*, leaf. Catkins nat. size; leaves ½ nat. size.

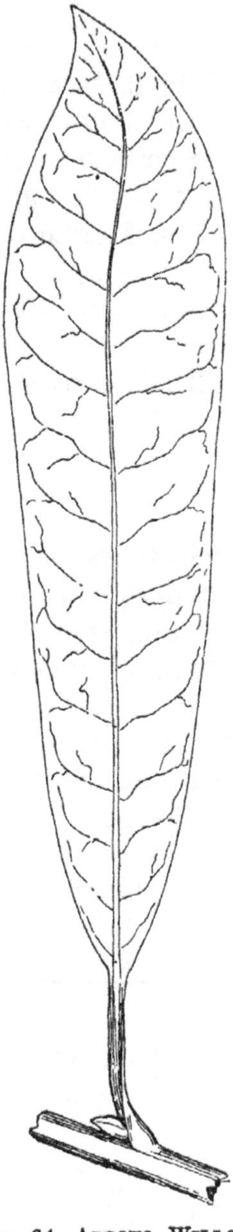
Fig. 64. Arroyo Willow (*Salix lasiolepis* Benth.). leaf, nat. size.

is the most characteristic willow in the beds of arroyos and gulches chiefly between 100 and 2500 feet altitude. It is a variable willow, yet its eccentricities are rather easily comprehended because very divergent states, especially leaf forms, may be collected from a single individual. Typically an inhabitant of the foothills it ranges occasionally to altitudes of 4000 feet.

Nuttall Willow

Fig 63 c, d, f

Nuttall Willow (Salix scouleriana Barr. var. flavescens Schn.) in California is usually a straggly shrub 2 to 8 feet high, or rarely a tree up to 25 feet high with a trunk ¾ to 1¼ feet in diameter. It inhabits moist north slopes or cupped hollows, or the vicinity of springs in the mountains. In the Coast Ranges it is limited to the vicinity of the sea.

In the Sierra Nevada and in the San Bernardino Mountains it occurs between 4000 and 10,000 feet. Beyond our borders it ranges east to the Rocky Mountains. In the Coast Ranges it is represented by the var. crassijulis Andr. and occurs only in the vicinity of the sea.

Velvet Willow

Velvet Willow (Salix sitchensis Sanson), often called Sitka Willow or Silky Willow, is a shrub 5 to 12 feet high or a tree up to 25 feet high, the trunk 2 to 10 inches in diameter. It is distributed along the California coast from the Santa Lucia and Santa Cruz mountains north to Marin and far north to Alaska. It also occurs in the Sierra Nevada on the west slope at 5000 to 7000 feet.

Fremont Cottonwood

Figs 65, 66

Fremont Cottonwood (Populus fremontii Wats.) sometimes called Common Cottonwood, is a handsome tree 40 to 90 feet high with ascending or wide-spreading branches forming a round-topped massive yellow-green crown supported on a short or long trunk 1

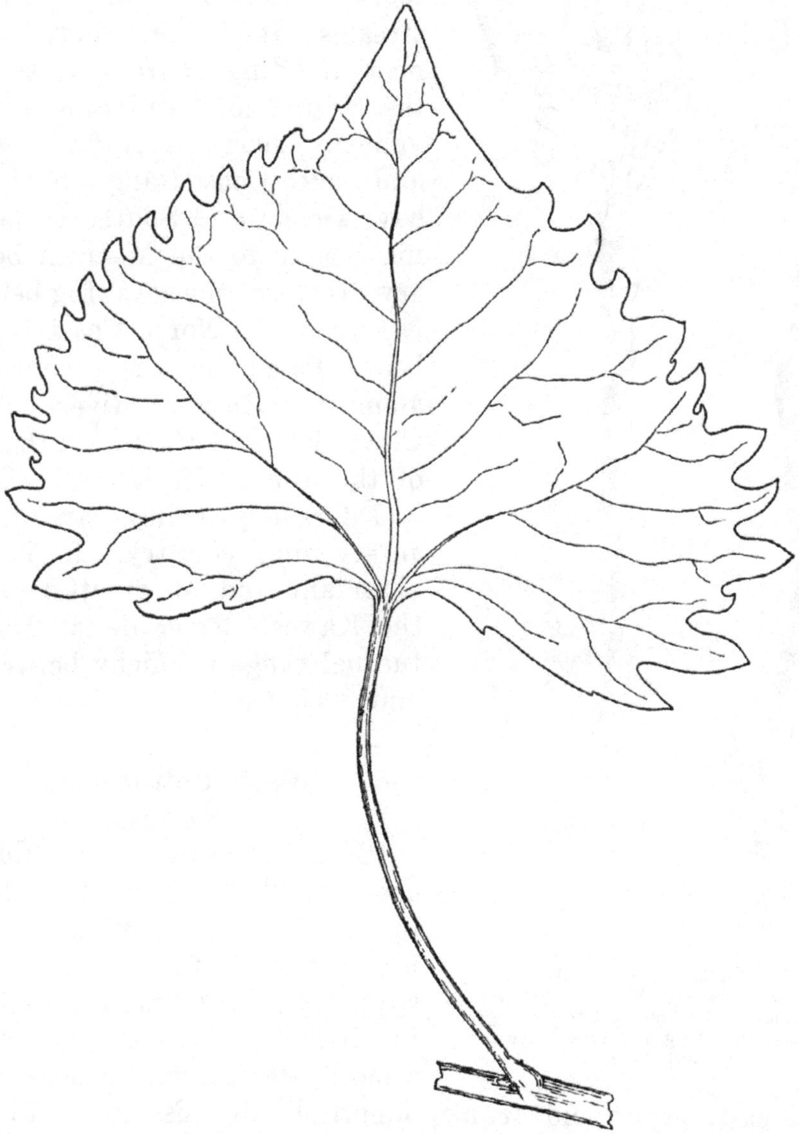

Fig. 65. FREMONT COTTONWOOD (*Populus fremontii* Wats). Typical leaf, nat size.

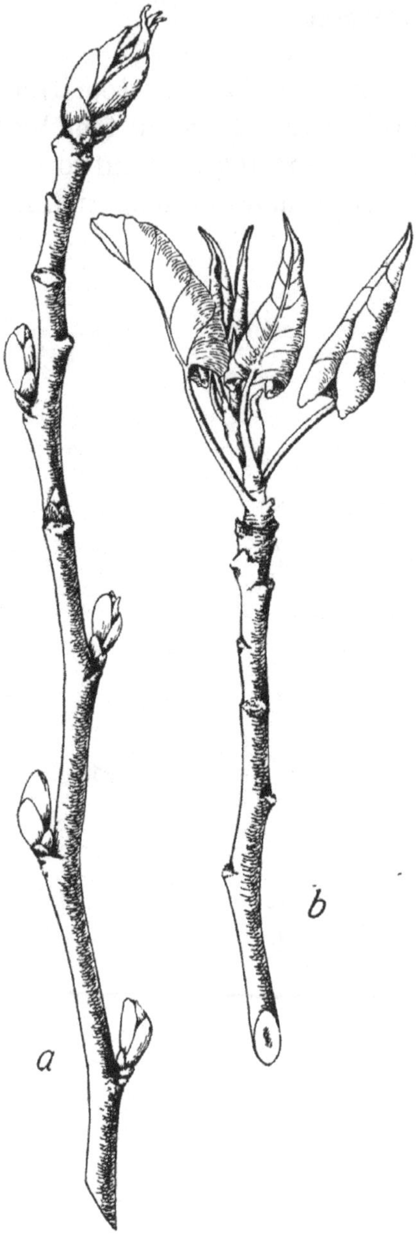

Fig. 66. FREMONT COTTONWOOD (*Populus fremontii* Wats). Winter branchlets with bursting buds, showing the unfolding involute leaves. ½ nat. size.

to 5 feet in diameter. The bark is white or whitish, on the trunk 1 to 5 inches thick and roughly fissured.

Populus fremontii inhabits stream beds and moist deltas in the valleys, rarely entering dry foothills except along living streams. It is distributed from near Redding southward through the Sacramento and San Joaquin valleys, Sierra Nevada foothills and South Coast Ranges to the Mohave Desert and Southern California nearly to the Mexican boundary. It shuns the coast fog belt and is rare in the North Coast Ranges, being noted by the writer only along the Russian River between Cloverdale and Ukiah, on the forks of the upper Eel River in Round and Gravelly valleys and in the intervening country. It is most abundant and of greatest size in the Kaweah River delta. Its altitudinal range is chiefly between 50 and 2000 feet.

Black Cottonwood

Fig 67a

Black Cottonwood (Populus trichocarpa T. & G.) is a tall tree, 40 to 125 feet high, with a rather broad crown of upright branches supported on a trunk 1 to 3 feet in diameter. The trunk bark is smooth, whitish with a usually yellowish cast, or on old trunks longitudinally fissured into long, narrow and rather smooth-surfaced dark plates.

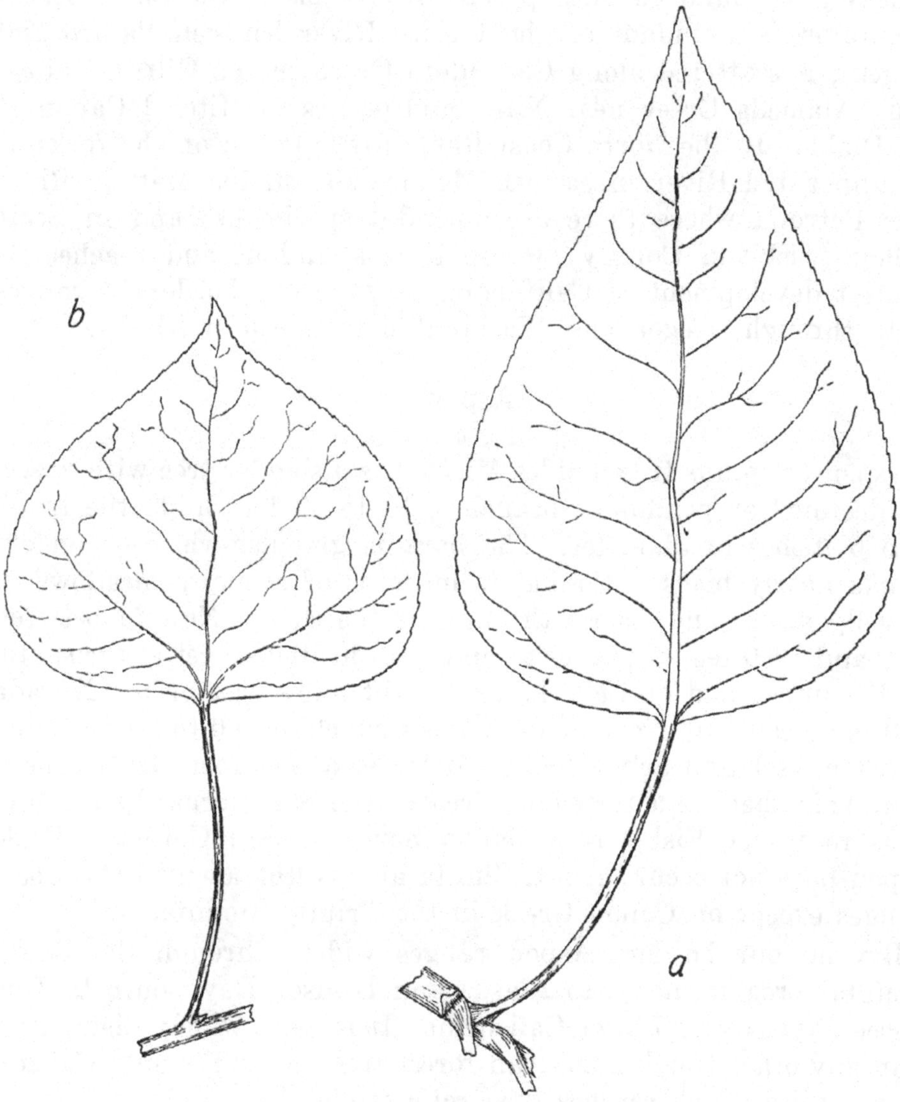

Fig 67. *a*, Black Cottonwood (*Populus trichocarpa* T & G), leaf, nat. size to ½ nat size. *b*, Aspen (*Populus tremuloides* Michx.), leaf, nat size.

Populus trichocarpa inhabits banks of valley or mountain streams or moist bottoms and is distributed through the Coast Ranges and Sierra Nevada, thence south to Southern California as far as Palomar Mountain (south limit). In the Sierra Nevada it occurs chiefly between 3000 and 6000 feet. There are fine trees on the floor of Yosemite Valley opposite Yosemite Falls, and in lower Hetch-Hetchy Valley, as well as in many similar places. In the Coast

Ranges it is found on most perennial streams in the Santa Lucia Mountains, is abundant on the Pajaro River between Pajaro and Sargent, is scattered along Carnadero Creek in the Gilroy Valley, along Alameda Creek near Niles, and occurs in Mitchel Cañon at Mt. Diablo. In the North Coast Ranges it is found on the forks of the upper Eel River in eastern Mendocino; on the Mattole River near Petrolia where there are splendid specimens; and in Scott Valley (Siskiyou County) where it is abundant and reaches its greatest development in California. Beyond our borders it ranges north through Oregon and Washington to southern Alaska.

Aspen

Fig 67b

Aspen (Populus tremuloides Michx.) is a slender tree with graceful declined or pendulous branches, 10 to 60 feet high, the trunk 3 to 6 inches in diameter. The bark is greenish-white or on old trunks nearly black. It inhabits margins of swampy meadows or gravelly slopes and occurs throughout the Sierra Nevada between 5000 and 8000 feet at the north and 6000 to 10,000 feet at the south. At the upper end of Lake Merced eight miles east of the Nevada Fall is a beautiful grove of these trees which are 60 to 80 feet high with trunks 1 to nearly 2 feet in diameter at 4½ feet. In Southern California there is a restricted area in the San Bernardino Mountains in upper Fish Creek Cañon north of San Gorgonio Peak. Aspen does not occur on Mt. Shasta and is unknown in the Coast Ranges except on Cañon Creek in the Trinity Mountains.

Beyond our borders Aspen ranges widely through the Rocky Mountain region, north to Alaska and Hudson Bay, south to Tennessee, Mexico and Lower California. It is more widely distributed than any other North American forest tree and is the only Californian species which reaches the arctic circle.

The wood is soft, whitish, fine-grained and fairly tough. It is too small a tree in California to be lumbered as it is occasionally in Oregon and Washington. Wherever abundant it is in demand for wood-pulp. With us it is used for fence-rails and also for fuel, since it burns when green and without sparks.

The Aspen, as an object in the landscape, is remarkable for its dancing leaves and their October hues. The petioles of the leaves are flattened vertically so that the equilibrium of the leaf is easily disturbed and even the slightest air-currents set the leaves a-quivering. In long dry falls the gorgeous golden colors of the foliage is one of the charms of the high Sierra forest.

California Laurel

Figs 68, 69

California Laurel (Umbellularia californica Nutt.), often called Bay Laurel and Pepperwood, assumes several distinct forms as modified by the character of the local habitat. In valley flats, cañon bottoms or on moist hill slopes it is a tree 25 to 100 feet high with a dense and often massive crown of long slender upright branches, the trunk 1 to 4 feet in diameter with a thin drab or brown bark ¼ to ½ inch thick. This is the most usual form, a type common in the Santa Cruz Mountains, Berkeley Hills, about Mt. Tamalpais, in the valley flats about Olema and in the hill country through the North Coast Ranges generally. The finest groves of Laurel in California were found on the river bench of the main Eel River near Camp Grant where there stood for several miles a pure stand of magnificent trees, until destroyed in recent years.

On the summit of the Berkeley Hills and in similar situations Laurel grows on rock outcropping and forms small many-stemmed clumps or bushy knobs which are a characteristic feature of such barren landscapes (Fig. 69). In the Coast Range chaparral it is sometimes seen as a low narrow-crowned dwarf 4 to 6 feet high, and on bluffs facing the sea its crown is developed in a contrary direction, spreading out over the ground as a low green mat of considerable diameter. Its most interesting modification occurs in wind-gaps in the hills of the San Francisco Bay region where it colonizes sharp north slopes in pure stands 5 to 20 feet high. These colonies are very dense, consisting of slender pole-like trees (usually crown-sprouts) with long trunks surmounted by a usually narrow broom-like top. The whole surface of the colony from above presents a very smooth and even appearance, as if clipped with a lawn mower, a feature which is due to control by the high wind velocities which prevail in the wind-gaps. The most beautiful and interesting example of this type occurs at Inverness on the south slope of the "first valley." Examples may also be seen in the cañon of the west branch of San Pablo Creek on the east slope of the Berkeley Hills near Fish Ranch.

Fig. 68. California Laurel (*Umbellularia californica* Nutt). Fruiting branchlet, nat. size.

The largest known California Laurel grows near the town of Cloverdale on an alluvial bench of the Russian River. Its crown is about 90 feet broad and 75 feet high and its perfect trunk is 4 feet 10 inches in diameter at 5 feet above the ground.

Umbellularia californica is widely distributed in both the Coast Ranges and Sierra Nevada, ranging south into Southern California and north to the South Fork of the Umpqua River in southern Oregon. To many travelers in central California it is one of the most pleasing of our species on account of its dense dark crowns which form a foil to dry brown hills of which this tree is often the only arboreal tenant.

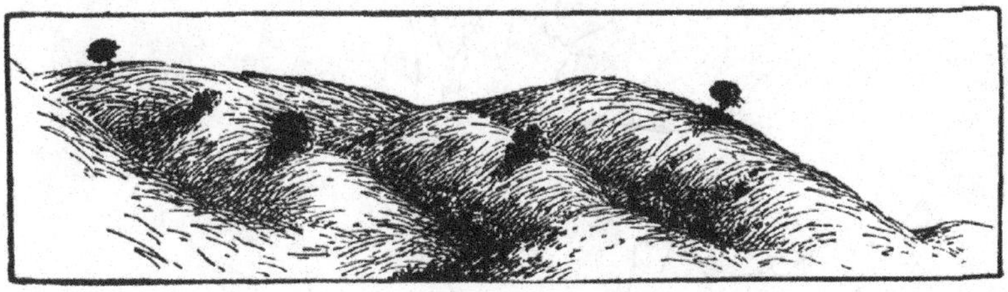

Fig 69. CALIFORNIA LAUREL, clumps (like round knobs) on rock outcroppings in the Berkeley Hills.

The wood is heavy, hard and strong and takes a high polish. Broad planks of it often show figured grain of great beauty. It has been used for furniture (especially bed-room sets), stave timber and shoe lasts.

Western Sycamore

Figs 70, 71

Western Sycamore (Platanus racemosa Nutt.) is a tree 40 to 90 feet high with a massive crown of wide-spreading limbs supported on a trunk 1 to 5 feet in diameter. The bark is smooth but exfoliates thin reddish-brown plates which expose greenish or whitish areas and give the trunk a mottled appearance.

Platanus racemosa is most characteristic of stream bottoms, either constant or summer-dry, in the more arid parts of California between the desert and the areas of high precipitation. It is most abundant in the South Coast Ranges but does not extend into the desert. On the other hand it does not enter the summer fog belt and it has never been seen in the North Coast Ranges. It ranges south through Southern California to Lower California and northward through the Sierra Nevada foothills and San Joaquin and Sacramento valleys as far as Tehama County.

Sycamore is often, especially in the South Coast Ranges and in Southern California, the only tree in its locality. Thin groves on alluvial benches or in river bottoms are remarkable for their leaning trunks which have diverged from the perpendicular by reason of the shifting nature of soil in stream beds. On account of its

Fig 70 Western Sycamore (*Platanus racemosa* Nutt) Branchlet with string of balls ⅓ nat size

large irregular crown, the long reach of its branches, its ample foliage turning bronze-color in late autumn, its leafy light-gray bark and handsome strings of balls, Sycamore holds the attention of the traveler in the characteristic country which it favors and takes a permanent place in his interest.

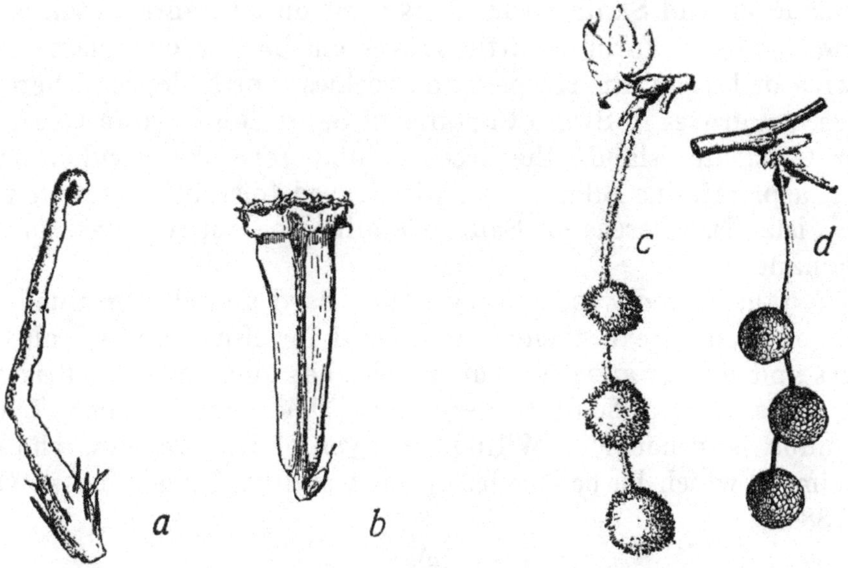

Fig. 71. **Western Sycamore**, capitate clusters of flowers spaced along a pendulous axis. *d*, Staminate inflorescence; *c*, pistillate inflorescence; *a*, pistil; b, stamen. *c* and *d*, nat. size; *a*, and *b*, 12 times enlarged.

Desert Mahogany

Desert Mahogany, often called Mountain Mahogany or Curl-leaf Mahogany (Cercocarpus ledifolius Nutt.), inhabits arid mountains. It is most common on the desert ranges of the Death Valley region, but is also common on the east slope of the Sierra Nevada, and on the desert slope of the San Bernardino and San Gabriel mountains. It occurs on Mt. Pinos and the mountains of Santa Barbara County, and is also found in the southern Sierra Nevada, the inner North Coast Ranges, and the Warner Mountains of Modoc County.

The wood is very dense, fine-grained and heavy; it has a reddish mahogany color, whence the common name, but splits badly in seasoning. It has been used extensively for fuel at mines in the deserts.

The Trask Mahogany (Cercocarpus traskae Eastw.) is a little-known tree 15 to 25 feet high growing on Santa Catalina Island where it was discovered in a steep cañon on the south side.

Catalina Ironwood

Catalina Ironwood (Lyonothamnus floribundus Gray) is a slender tree 20 to 55 feet high, the trunk ½ to 1 foot in diameter. The leaves are linear and somewhat Oleander-like. It is an insular species found only on the islands of Santa Catalina, San Clemente,

Santa Rosa, and Santa Cruz. It is common on Santa Catalina but grows by itself, forming little groves one-half to one-quarter acre in area or less, on the steepest and rockiest north slopes. There are three fine groves in Swain Cañon. On Santa Rosa, Santa Cruz, and San Clemente islands the trees exhibit fern-like divided leaves (var. asplenifolius Bdg.), the entire-leaved form being absent from these islands, whereas on Santa Catalina the entire-leaved form is dominant.

The pinkish red wood is very heavy, hard and close-grained. On account of its great strength it is used for fish-poles and also for canes and similar articles. This species has long been in cultivation but the seed is difficult to germinate. The generic name, Lyonothamnus, is in honor of William S. Lyon of Los Angeles, who sent specimens which he had collected on Catalina Island to Asa Gray in 1884.

Islay

Islay (Prunus ilicifolia Walp.) is an evergreen tree with spiny-toothed leaves and broad rounded crown 15 to 40 feet high, which inhabits valleys or cañon bottoms from the Napa Mountains to San Fernando Valley and southward to Lower California. It also occurs as a small rigid bush and is, in this form, a distinctive element in the composition of the chaparral from Monterey County to Southern California.

The dense racemes of flowers mature 2 or 3 large red or dark purple drupes with very large stone. The flesh is agreeable and enticing, but to the inexperienced traveler surprisingly thin. Island Cherry (Var. integrifolia Sudw.) is the entire-leaved form on Catalina Island.

Bitter Cherry (Prunus emarginata Walp.) forms extensive thickets in the high mountains from 4000 to 8000 feet, although along the coast it occurs from 500 to 1500 feet. On the lower borders of its altitudinal range it sometimes becomes a slender tree 15 to 25 feet high. Its leaves are deciduous and its flowers are in corymbs.

Honey Mesquite

Fig. 120

Mesquite, more specifically the Honey Mesquite (Prosopis juliflora DC.), called Algaroba by the Mexicans, is a small tree, the short trunk dividing into wide-spreading branches forming a rounded or low broad crown commonly 8 to 20 feet high and usually twice or thrice as broad. In well-developed individuals the ends of the outer branches usually rest on the ground. It is a characteristic species of the Mohave and Colorado deserts, ranging northward to Death Valley and into the upper San Joaquin Valley in Kern County, eastward to Texas and southward in various forms to Chile.

It is a remarkable desert tree and is exceedingly useful in many ways to the desert tribes and white settlers. The pods furnish a staple food to the Indians as well as to their saddle ponies. The wood is used by Indians for building houses and for household implements. In certain situations wind-blown desert sands gradually bury the trees, save the tops, and such mounds are excavated for fuel —one of the great resources of miners and settlers in the desert region. The flowers furnish food to the honey bee, whence the common name, Honey Mesquite.

Screw-bean Mesquite

Fig 121

Screw-bean Mesquite (Prosopis pubescens Benth.), the Tornillo of the Mexicans, is an erect shrub or small tree 10 to 25 feet high with spiny branches and trunk 3 to 10 inches in diameter. It grows in sandy or gravelly washes or ravines and is distributed throughout the Colorado and Mohave deserts of Southern California, northward to Death Valley, eastward to southern Utah and New Mexico, and southward to northern Mexico. The tightly coiled pods or beans are remarkable structures botanically. They are also of economic interest, since being sweet and nutritious, they are used as food by the Indians and as fodder by cattle.

Palo Verde

Fig 122

Palo Verde (Cercidium torreyanum Sarg.) is a small but very interesting tree 15 to 20 feet high with short trunk, smooth green bark and crown leafless for most of the year. It inhabits sandy washes or depressions in the Colorado Desert of Southern California, and ranges eastward into southern Arizona and southward into Lower California and Sonora. The leaves fall soon after they appear in March, but the trees still present a cheerful appearance on account of the bright green bark (whence the Spanish name), which is all the more pleasing on account of the contrast with the parched desert scenery. The pods fall in July and are harvested by the native tribes of the region who prepare them for food. The branchlets are browsed by cattle, horses and deer, and the flowers are visited by the honey-bee.

Smoke-Bush

Smoke-bush (Parosela spinosa Hel.) is a very spinose and nearly leafless ashy-gray low shrub or small tree 4 to 25 feet high with intricately much-branched top and trunk 2 to 10 inches in diameter. It is common in dry washes from Palm Springs and the Chuckawalla Bench eastward throughout the Colorado Desert to the Gila River in Arizona, and southward to San Felipe, Sonora and Lower California. It has been so named on account of its appearance, being so truly deceptive as to cause the uninitiated in the desert to watch it with speculative wonder as if it were a column of smoke.

Desert Ironwood

Desert Ironwood (Olneya tesota Gray), or Arbol del Hierro of the Mexicans, is a spreading tree 15 to 20 feet high with short trunk ¼ to 1½ feet in diameter. It grows in the desert valleys of Southern California from San Felipe northward to Chuckawalla and Indio, eastward to Arizona, and southward into adjacent parts of Mexico. Its wood is remarkably hard and heavy and is used by desert Indians for arrow parts and tool-handles.

California Buckeye

Figs. 72, 73, 74, 75, 76, 77, 78

California Buckeye (Aesculus californica Nutt.) is a tree 10 to 30 feet high with a low broad rounded crown and trunk ⅓ to 3 feet in diameter. The bark is smooth and white, or on old trunks fissured into thinnish scaly plates. It inhabits valley flats, river bottoms or more particularly lower hill slopes. In bowl-like hollows in the hills, at the heads of cañons, or at the edge of chaparral it often forms open thickets of many-stemmed shrubs 5 to 8 feet high.

Aesculus californica is widely distributed in the Coast Ranges and Sierra Nevada foothills, and extends north to South Fork Trinity River and Redding, and southward to Fort Tejon and Antelope Valley. More characteristic of the interior hills, the Buckeye in a few places ranges to the immediate coast although shunning

Fig 72 Buckeye (*Aesculus californica* Nutt). Flowering branch, ⅓ nat. size.

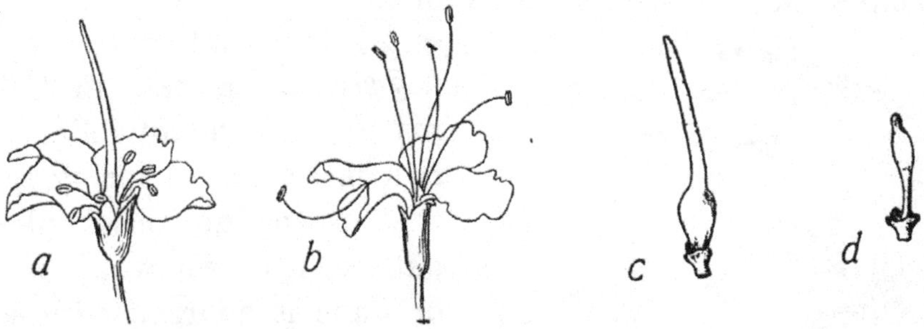

Fig 73. Buckeye *a*, Perfect flower, long-styled; *b*, staminate flower with short-styled sterile ovary; *c*, fertile pistil, ovary sessile, *d*, sterile pistil, ovary stipitate. nat. size

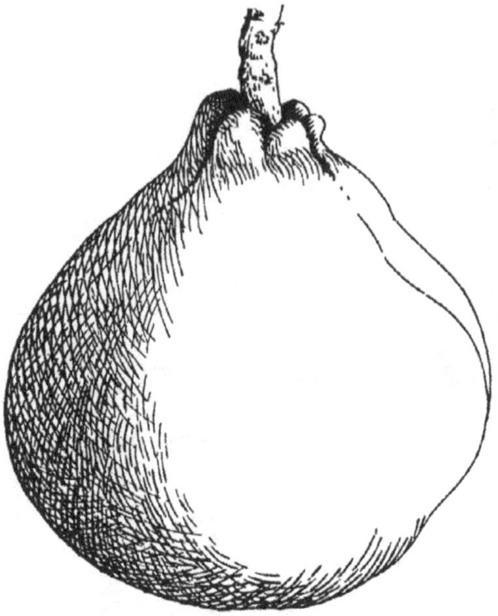

Fig 74. BUCKEYE pod. nat. size.

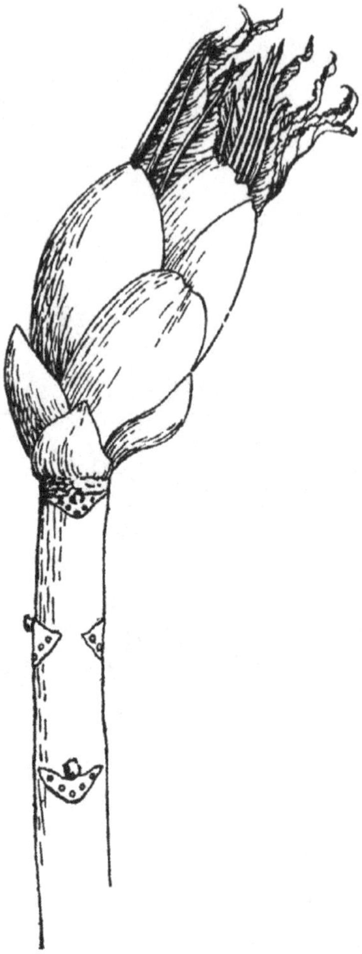

Fig. 76. Bursting winter bud of BUCKEYE, the leaf scars of last summer's leaves showing below nat. size.

the main Redwood belt. The trees are usually scattered along gulches or water-courses in the hills or form open groves on the lower slopes of the foothills just above the valley level, sometimes occurring in large size on river benches or moist flats.

The California Buckeye, a tree of northern origin, with large winter buds and ample leaf surface, is rather interesting in its adaptation to arid dry-season habitats in California. The abundant foliage, of a rich dark green in spring, rounds the crown into most pleasing outline, a crown which in May or June is adorned with showy clusters of flowers. By early July the foliage is brown and heat-crumpled, and the leaflets are

Fig 75. BUCKEYE seed. nat. size.

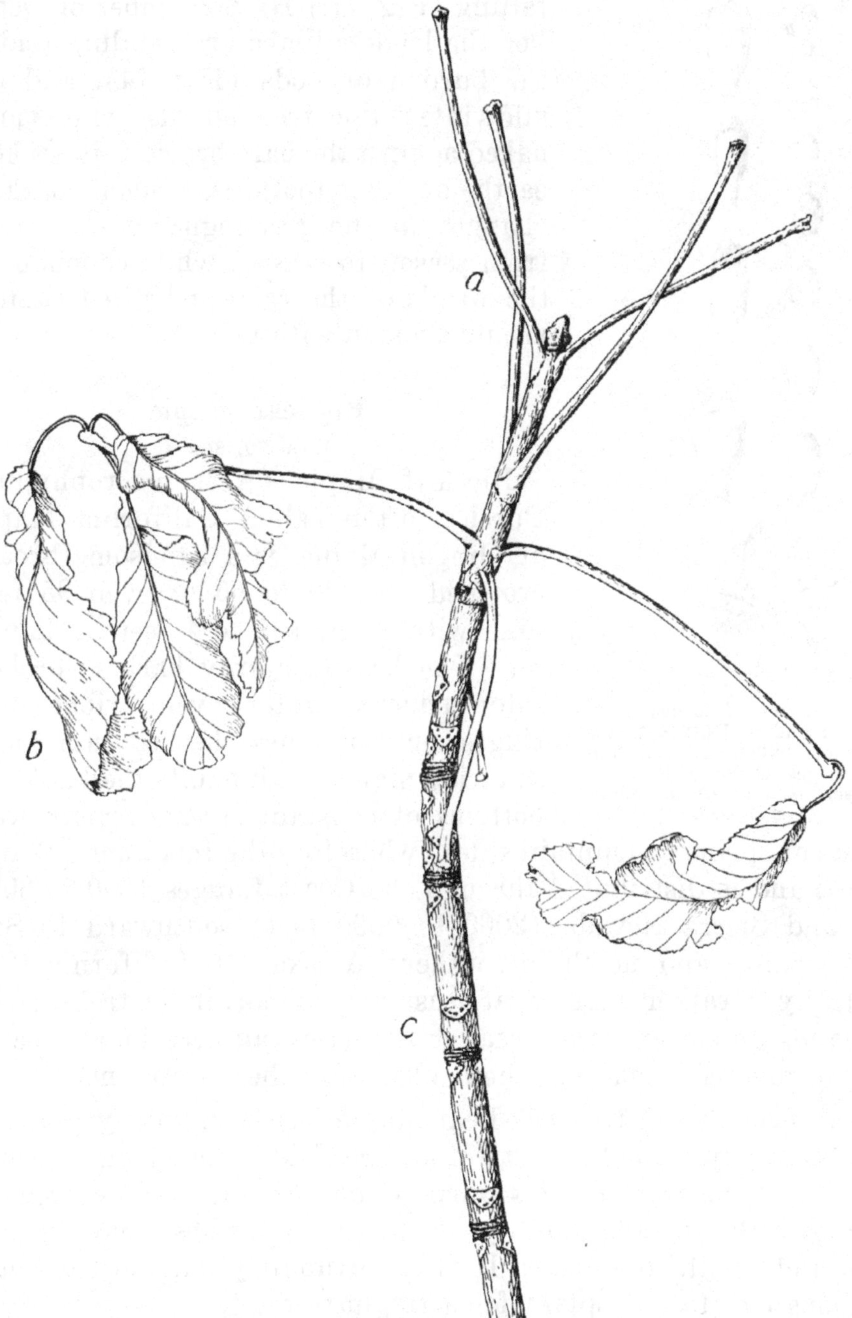

Fig. 77. BUCKEYE, September branchlet. *a*, Petioles of compound leaf from which the leaflets have fallen (below are petioles turned downward); *b*, three of the five leaflets still persisting; *c*, triangular leaf-scars, marking the place of leaves in previous years, also dark circular scars indicating the position of the terminal winter bud at the end of each season's growth. ½ nat. size.

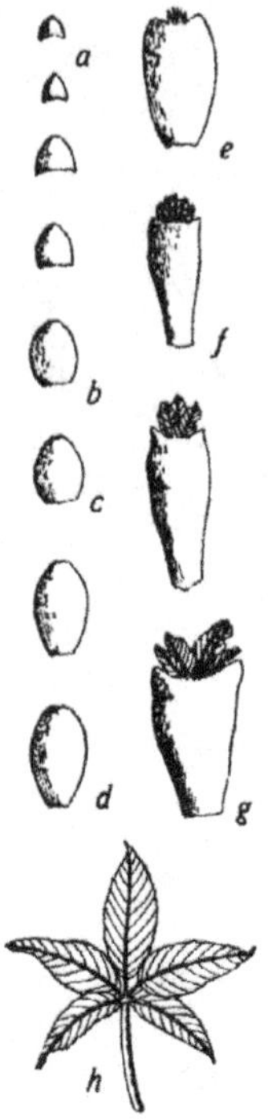

Fig 78 Series of scales from the winter bud of BUCKEYE, the vestiges of leaflets at the tops of the scales showing that the scale itself is a modified petiole ¼ nat. size.

falling (Fig. 77.) By September or October the leafless limbs are bending under the burden of pods (Fig. 74), and by midwinter the tree stands white and naked against the bare brown rain-soaked earth of the foothills. Such marked changes in the physiognomy of a tree from season to season, while common in the woods of the eastern United States, are uncommon with us.

Big-leaf Maple

Figs 79, 80a

Big-leaf Maple (Acer macrophyllum Pursh), often called California Maple or Oregon Maple, is a handsome broad-crowned tree 30 to 65 or even 95 feet high with a trunk 1 to 4 feet in diameter. The brownish-gray bark is broken into narrow interwoven ribbon-like ridges, or sometimes checked into small squarish plates. It inhabits the banks or bottoms of constant streams, moist valley flats or springy mountain sides (wherefore the folk-name, Water Maple), and is distributed through the Coast Ranges (200 to 3000 feet) and Sierra Nevada (2000 to 5000 feet) southward to San Diego County and north to southern Alaska. In California it is forestrally a rather rare or at least very subordinate tree in our woodlands and over large areas it is merely an occasional tree in the most favored situations, the banks or benches of streams.

As an ornamental tree Big-Leaf Maple has been widely planted. It is also highly valued as a street and roadside tree, being the most available native tree for this purpose, on account of its extremely rapid growth, its adaptability to street conditions, and its fine crown and beautiful foliage. It is unfortunately our most difficult deciduous tree to transplant from the nursery.

The wood is reddish-brown with white sapwood. It is rather hard and close-grained, takes a high polish and works easily. It has been used for tool-handles, furniture and interior finish. Moun-

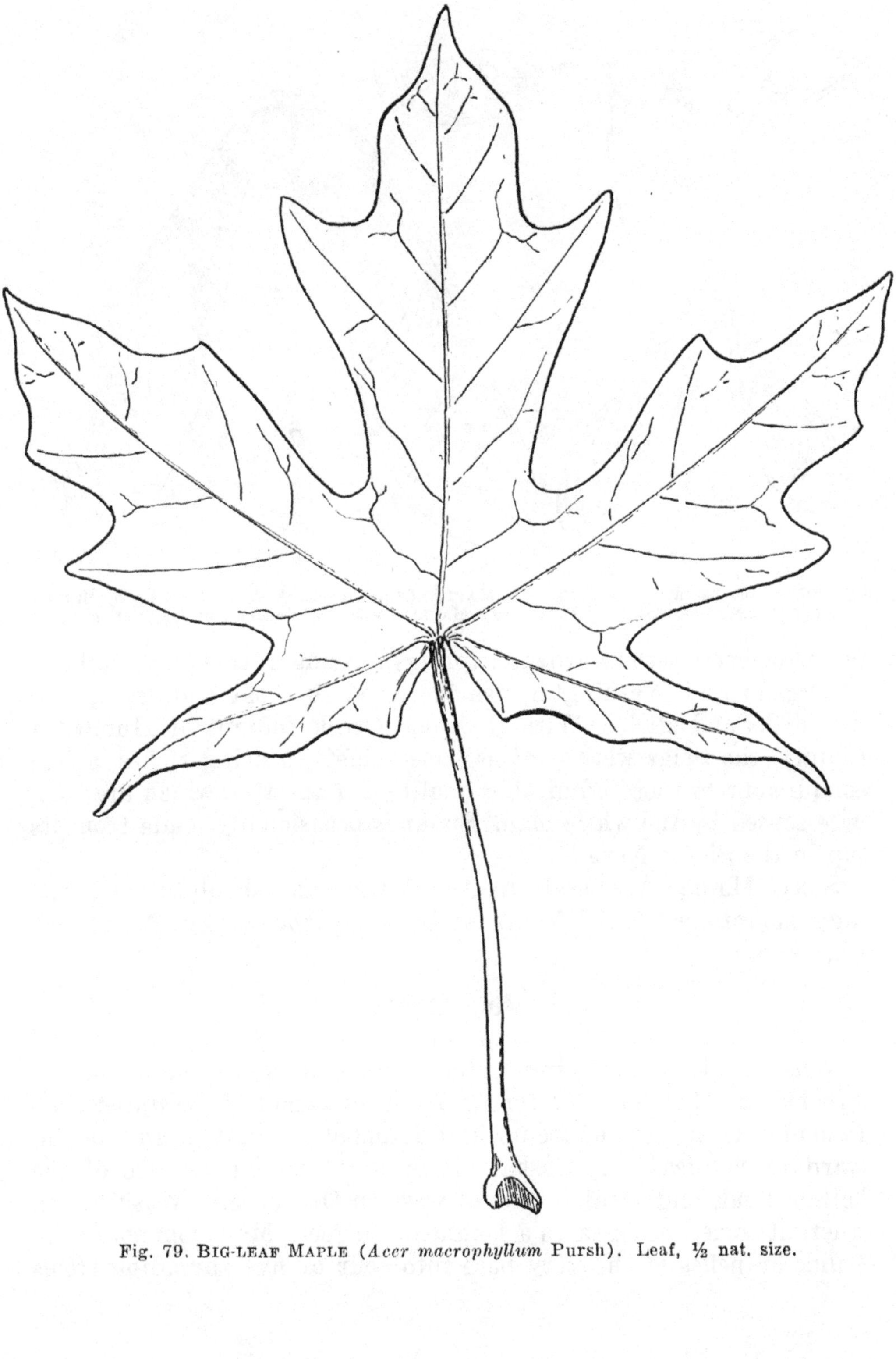

Fig. 79. BIG-LEAF MAPLE (*Acer macrophyllum* Pursh). Leaf, ½ nat. size.

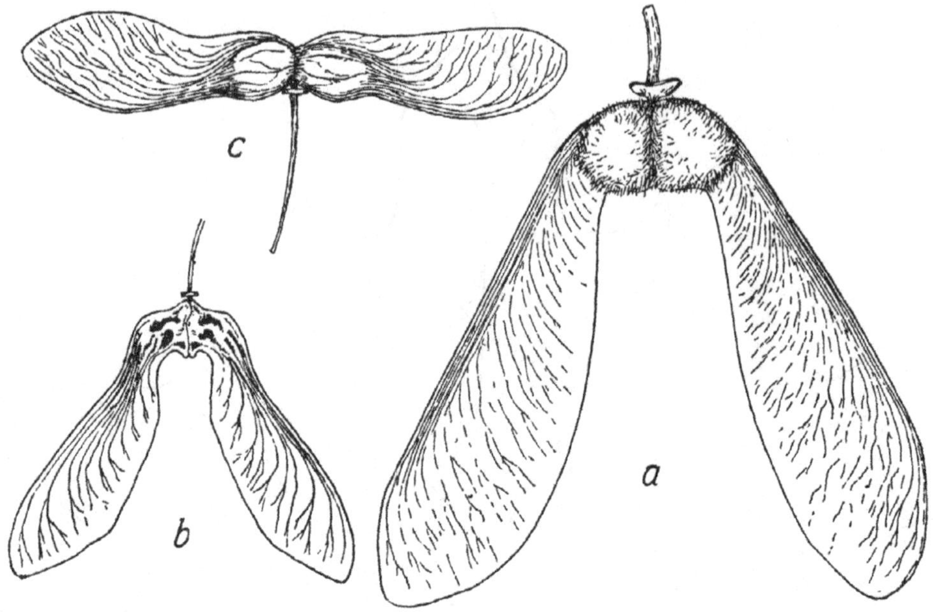

Fig 80 Maple samaras *a*, BIG-LEAF MAPLE (*Acer macrophyllum* Pursh). *b*, SIERRA MAPLE (*Acer glabrum* Torr). *c*, VINE MAPLE (*Acer circinatum* Pursh). nat. size.

taineers choose second-growth saplings for single-trees. The Indians of Oregon and Washington use the wood for boat paddles, gambling disks and sticks. With Redwood Creek Indians of Humboldt County, the twigs were used as "medicine" to bring riches, a fine compliment to those admirable qualities of the tree which are likewise sensed by the white man. Sugar is occasionally made from its sap in the Sierra Nevada.

Sierra Maple (Acer glabrum Torr.) is a small shrub found in the high mountains of California and east to the Rocky Mountains. Fig. 80 b.

Vine Maple

Figs 80c, 81

Vine Maple (Acer circinatum Pursh) inhabits the banks of streams and the depths of forests from the cañon of the upper Sacramento River to Mendocino and Humboldt counties and northward to southeastern Alaska. It is an attractive feature of the bottom lands and great forests of western Oregon and Washington, where it sometimes attains a height of 35 feet. Most commonly the trunk branches at the very base into four or five spreading stems

Fig 81. Vine Maple (*Acer circinatum* Pursh). Leaf. nat. size.

which curve over and, touching the ground, take root. Further offsets arise and result in extensive and well-nigh impenetrable thickets.

California Box-Elder

Fig 82

Box-Elder (Acer negundo var. californicum Sarg.) is a broad-crowned tree 30 to 45 feet high with trunk 1 to 1½ feet in diameter. It inhabits banks or bottoms of constant streams and is distributed through the Coast Ranges, Sacramento and San Joaquin valleys and Sierra Nevada foothills, ranging south to the San Bernardino and San Jacinto mountains. The trees in a locality are usually few and scattered and there are wide gaps in its distribution. It is rather more common along the Pajaro River between Pajaro and

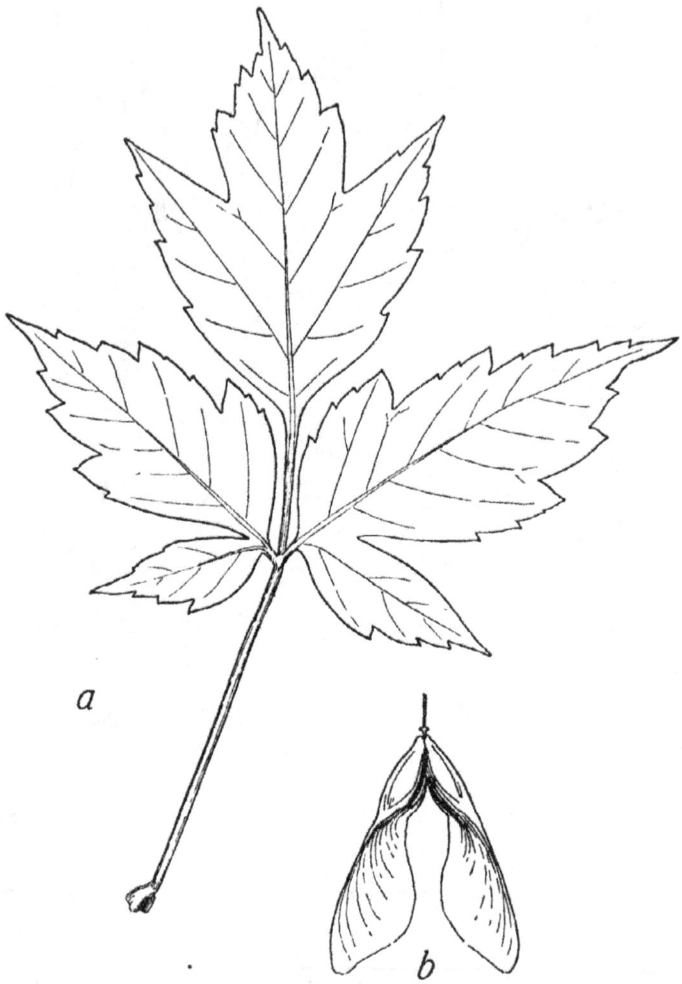

FIG 82 CALIFORNIA BOX ELDER (*Acer negundo* var *californicum* Sarg). *a*, Leaf, ½ nat size; *b*, samara, nat size

Sargent than seen elsewhere by the writer. On account of its hardiness and attractive foliage it is frequently used as a street tree in the cities and towns of California.

Mountain Dogwood

FIG. 83

Mountain Dogwood (Cornus nuttallii Aud.) is a slender tree 15 to 50 feet high with irregular crown and smooth whitish bark. It inhabits cool depths of mountain forests where the shade is

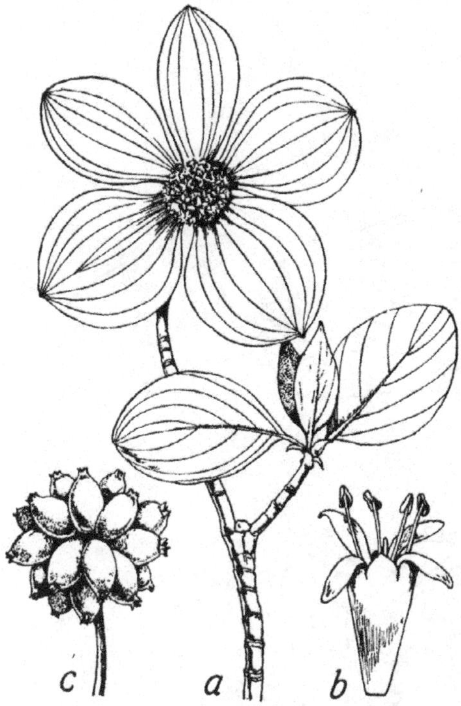

Fig. 83. Mountain Dogwood (*Cornus nuttallii* Aud.). *a*, Fl. branchlet, ⅓ nat. size; *b*, flower, 3 times nat. size; *c*, fruiting cluster, ½ nat. size.

deep and the soil moist. The individuals are usually scattered or rarely in small clusters. In the Sierra Nevada it occurs between 2500 and 5000 feet; in the Coast Ranges it is limited to the seaward ranges and to the middle North Coast Ranges. It extends south to the San Bernardino and San Jacinto mountains of Southern California and north to British Columbia.

The white bracts which surround the cluster of small flowers are showy and are produced in early spring. The tree in full flower never goes unnoticed by the traveler; its slender habit and the freshness of its bloom impart to the forest a sense of lightness and animation possessed by no other tree in the Pacific woods.

Madroño

Fig 84

Madroño is a genuine Arbutus, the Arbutus menziesii Pursh of botanists. It inhabits rich slopes or sometimes rocky spots in the foothills or mountains or gravelly valleys, and is distributed from Southern California north through the Coast Ranges to British Columbia. It occurs in the Sierra Nevada from the Tehama County foothills to the South Fork Tuolumne River, but is not common. It is most common in the North Coast Ranges where it grows on mountain ridges, slopes and gravelly valleys, reaching its greatest development in Mendocino and Humboldt counties where, as an associate of Tan Oak, Douglas Fir and Black Oak, it is everywhere a striking feature of the woods in the ''Bald Hills'' region inside the Redwood belt.

Madroño is rarely symmetrical and the older the tree the more unsymmetrical as a rule. This is notably the case in the Mendocino and Humboldt woods, where it is invariably pushed to one side

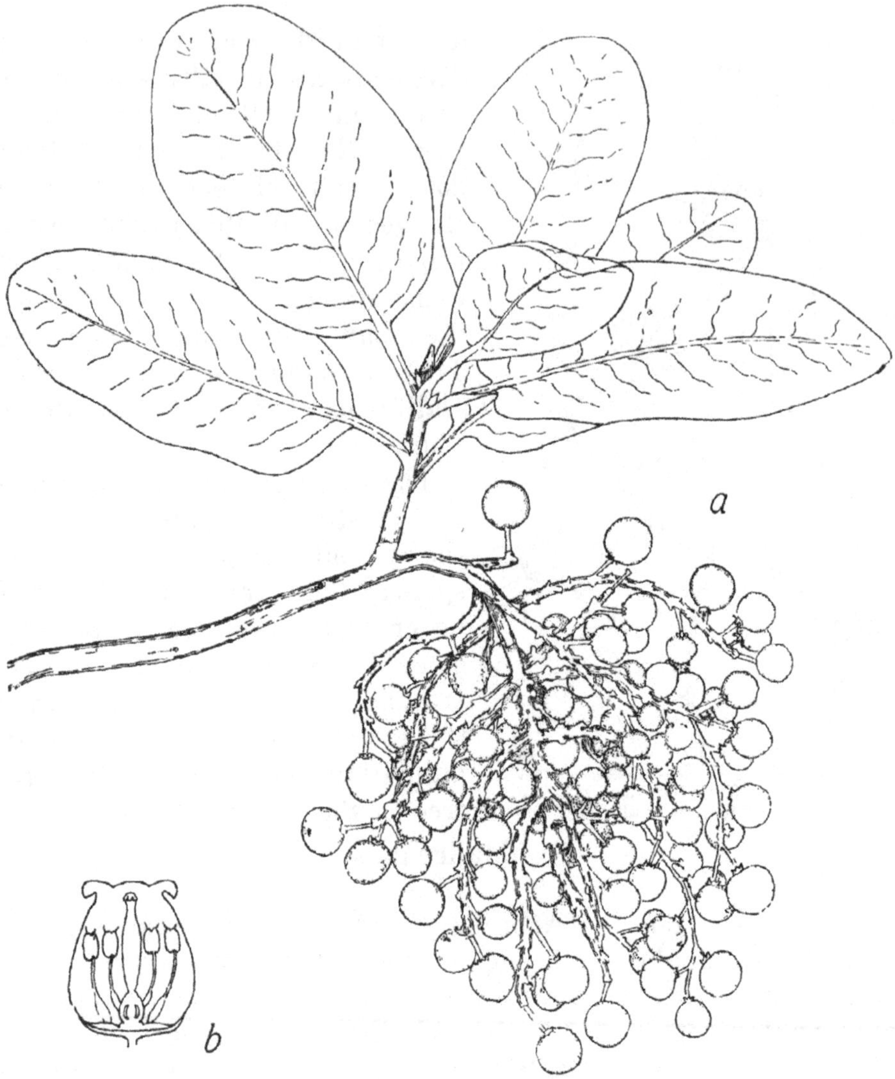

Fig. 84 MADRONO (*Arbutus menziesii* Pursh). *a*, Fruiting branchlets with berries, ½ nat. size; *b*, longitudinal section of flower, 2½ times nat. size

when in light competition with Douglas Fir or Tan Oak. Huge Madroño crowns, wholly one-sided, are frequently met with; sometimes the aggressive companion trees disappear and leave these irregular Madroños standing alone. Very frequently one finds a long trunk curving out of the perpendicular 20 or 30 feet and up 60 or 70 feet to a wisp of a crown occupying a very small area of the forest canopy. Such trees are remarkable for their curving and often huge trunks, which are commonly very tall and often flattened contrary to the direction of curve.

The contrast of color in bark and foliage is the most striking feature of the tree to the traveler. On branches or young trunks the bark is deep red and very smooth. When the summer growth begins it is deciduous in thin layers, revealing a satiny ground of pale green which ages into the characteristic red or terra-cotta. The bark parts very readily from the wood, and girdled trunks, girdled without apparent provocation or as mere pastime, are often seen in the forest. Their color and smoothness, so attractive and distinctive to the forest lover, seem but as a challenge to the destructive instincts of the gross. Fortunately the trunk has the peculiarity of retaining a thin inner layer of the bark which the vandal often ignorantly overlooks and the tree may live and flourish in spite of mutilation.

In its native woods Madroño, as a species of unusual beauty, has been little regarded by Californians, except by a chosen few who know the northern ranges. To these, what wonder that this tree inspires both pen and brush! It has been well depicted on the canvases of Welch and other artists and has, in addition a permanent place in our California literature. When a boy Francis Bret Harte journeyed overland from Eureka to the Bay of San Francisco and in early youth gave to Californian poesy his happy lyric on this sylvan masquerader with velvet mantle and scarlet hose. So it is that the Madroño, the most handsome tree of Alta California—which enlivens the forests and the groves with its unrivaled woodland colors—will ever be associated with the man who has best portrayed the atmosphere of the Californian valleys and the foothills with their yellow cloaks.

No other of our trees, to those who know it in its regions of finest development, makes so strong an appeal to man's imagination—to his love of color, of joyful bearing, of sense of magic, of surprise and change. He walks the woods in June or July and rustles the mass of gold-brown leaves fresh fallen under foot, or rides for hundreds of miles across the Mendocino ranges—and always with a sense of fresh interest and stimulation at the varying presence of this tree.

Although of slight economic importance as a timber species, it is in every other way a notable tree. Its crown of flowers and masses of crimson berries, its burnished foliage and terra-cotta bark, its manner of branching and habits of growth are alike full of interest and of charm. Wherever it grows the traveler, the forester, the hunter, the artist or the botanist is held by its spell, and none such worthy of the name ever came out of the northern woods but returned to them again and again in waking or in dreaming moments, guided by the ordered paths of the intellect or loitering free in the crimson uplands of the imagination.

Oregon Ash

Fig. 124

Oregon Ash (Fraxinus oregona Nutt.) is a tree 30 to 80 feet high with a rather broad round-topped crown and trunk ½ to 3 feet in diameter. The trunk bark is gray-brown, 1/6 to ¾ inch thick, fissured into narrow freely interlaced ridges.

Fraxinus oregona inhabits stream banks in ravines, river bottoms, and moist flats in valleys. It is widely distributed through the Sierra Nevada and Coast Ranges and is occasional in the higher ranges of Southern California south to San Diego County. Beyond our borders it extends northward through western Oregon (where it is most abundant) and Washington to British Columbia. It grows along the Sacramento and San Joaquin rivers, ranging into the Sierra Nevada foothills to altitudes of 1000 to 2000 feet. In the South Coast Ranges it occurs only at scattered stations but is more common and widely distributed in the North Coast Ranges, especially in Lake, Mendocino and Humboldt counties, where it is sometimes the dominant tree in the delta swamps of mountain valleys.

Its wood is rather coarse-grained, hard and strong and is used for interior finish, furniture, wagon parts, and implement handles. The supply in California is too small to be of importance other than for local uses.

Arizona Ash

Arizona Ash (Fraxinus velutina Torr.) is a tree 15 to 30 feet high with grayish somewhat fissured bark and willow-like leaflets. It inhabits the banks of streams in cañons or the borders of lakes or springs at altitudes of 2200 to 4000 feet. It is distributed from Owens Lake and the Panamint Range to the southwest base of Mt. San Jacinto and southeastern San Diego County, and ranges east through southern Nevada to western Texas.

Leather-leaf Ash (var. coriacea Jepson) is a tree 20 to 30 feet high with round-topped crown and rough gray trunk bark. It inhabits desert regions and is found on the east side of Owens Lake (4000 feet) and at the east base of Mt. San Jacinto (2000 feet). It ranges east to southern Utah.

Dwarf Ash (Fraxinus anomala Wats.) is a tree 15 to 20 feet high or a low spreading shrub, growing in desert washes or borders of desert streams. It occurs in the Providence Mountains and ranges east to Arizona and Colorado.

Desert Willow

Desert Willow (Chilopsis saligna Don) grows along water-courses or washes in the Mohave and Colorado deserts of Southern California, recurring locally in the San Jacinto Valley. Its common name is derived from its narrow willow-like leaves, although of course it is not in any way related to the willows. The large 2-lipped corollas, which cover the crown with pinkish bloom in April, and the tufted seeds, refer it unmistakably to the Bignonia Family. Its peculiar habit and showy flowers make it an interesting feature of the desert flora.

Blue Elder

Fig 85

Blue Elder (Sambucus glauca Nutt.) is a tree 15 to 28 feet high with roundish or irregular crown and trunk ½ to 1½ feet in diameter, or most commonly a roughish bush 5 to 10 feet high with several to many upright main stems. It ranges from Washington to Southern California, occurring in both the Sierra Nevada and Coast Ranges.

The bush-like clusters, which are common both in the valleys and mountains, may aspire to develop a single tree-like trunk, but rarely do so since the truly arboreous form is forestrally rare, such individuals being widely scattered or occurring in small clusters.

In late summer or winter Sambucus glauca is scraggly and unattractive on account of its habit of dying back. Its best season is early spring when the clumps round out full-foliaged heads which are truly attractive, as they are also a few months later when dowered with flowers.

The berries, often produced in great abundance, are used in rural cookery. Multitudinous jays, woodpeckers and other birds feed voraciously upon them and so distribute the seeds, especially along fence lines in the valleys.

FIG. 85. BLUE ELDERBERRY (*Sambucus glauca* Nutt.). *a*, Flowering branch, ½ nat size; *b*, flower from above, 9 times nat. size.

GEOGRAPHIC DISTRIBUTION OF THE CALIFORNIA SILVA

Peculiar and Characteristic Species

The number of species of forest trees in California which may properly be called such is 92. Of the total 39 are coniferous trees, 3 are palm or palm-like trees, and 50 are broad-leaved trees. The most interesting and striking features of the silva of California relate to its composition, the geographical distribution of the species and their biological history.

California is remarkable for its development of coniferae, not only in number of species (which exceeds that of any other equal area), but in size of the individual trees and their forestral development. This statement is peculiarly true of the true pines of which we have 15 species. The oaks may well be contrasted with the pines. Of true oaks we have 15 species, as many as the pines, but their forestral development is comparatively insignificant, except for the Tan Oak, and their biological interest is considerably less.

The silva of California is remarkable for the number of species peculiar to California or which here attain their greatest development. It is also interesting for the number of extremely local species—species which are confined to a single or few localities or have a restricted range.*

The number of species strictly peculiar to California is 20, a relatively large number. The number of species which may be called typically Californian is also large, namely 52. By typically Californian is meant that a species has here its greatest development and is restricted to our area, or ranges beyond it no great distance or only in a feeble or uncertain manner.

The distinctive character of the silva of California is emphasized by a comparison of it with the silva of the eastern United States. These silvas have only two species in common, namely the Aspen and the Black Willow. The California silva has, however, marked relationships with the silva of Oregon and Washington and in a less degree with the silva of the Rocky Mountains.

Nearly all our species of Pinaceae and Cupressaceae are peculiar to the Pacific Coast. Although about 13 species range east to the

---

*See sections on "Tree Islands" and the "Klamath Mountains."

Rocky Mountains only a few of these have any considerable development or extension in that region. All of our species of Fagaceae are peculiar to the Pacific Coast save one only, Quercus chrysolepis Liebm., which ranges east in a limited manner through Arizona to New Mexico.

In the following list of typically Californian species, the species strictly peculiar to the State are marked with an asterisk.

TYPICALLY CALIFORNIAN TREES.[4]

Sugar Pine.	*Mendocino Cypress.	Giant Chinquapin.
*Foxtail Pine.	*Sargent Cypress.	Tan Oak.
Yellow Pine.	*McNab Cypress.	*Valley Oak.
Big-cone Pine.	*Piute Cypress.	*Blue Oak.
*Digger Pine.	Sierra Juniper.	Mesa Oak.
*Torrey Pine.	*California Juniper.	Maul Oak.
Knob-cone Pine.	*California Nutmeg.	Coast Live Oak.
Monterey Pine.	*California Fan Palm.	Interior Live Oak.
Bishop Pine.	Joshua Tree.	Black Oak.
Weeping Spruce.	Mohave Yucca.	*Oracle Oak.
Big-cone Spruce.	Red Willow.	California Laurel.
Red Fir.	Yellow Willow.	*California Buckeye.
White Fir.	Arroyo Willow.	Western Sycamore.
*Santa Lucia Fir.	Fremont Cottonwood.	Islay.
*Big Tree.	*California Black Walnut.	*Catalina Ironwood.
Redwood.	*S. Cal. Black Walnut.	Madroño.
Incense Cedar.	White Alder.	Blue Elderberry.
*Monterey Cypress.	Red Alder.	

Life Zones in California

The geographic distribution of living forms is governed by certain definite factors, such as rainfall or humidity, temperature, altitude and insolation. Through the varying combinations of these factors living things are distributed into areas known as Life Zones, which differ in altitude and are characterized by certain means of temperature and averages of precipitation. Five zones are recognizable in California. The limits of the zones are in some cases sharply defined, while in other cases they are less definite, yet on the whole they serve to mark off in a broad way a segregation of living forms which may be observed in the field more or less clearly and which are indicated by certain dominant species called index species.

Of these zones, the Sonoran Zone is the lowest, with an average altitude from sea-level to 2000 feet. The mean annual temperature is 60 to 75 degrees, the rainfall is 1 to 15 inches. The Colorado and

[4] Compare special description of these species: see index.

Mohave deserts fall within the Sonoran Zone and are specifically known as the Desert Sonoran or Lower Sonoran. The most characteristic tree species are the Fan Palm, Palo Verde, Desert Ironwood, Honey Mesquite, Screw-bean Mesquite and Tree Yucca. Most of the areas of the San Joaquin and Sacramento valleys and the interior valleys of the Coast Ranges come within the Sonoran Zone, such areas being here called Valley Sonoran. The index species of trees for the Valley Sonoran are the Western Plane, Coast Live Oak, Valley Oak and Interior Live Oak. The interior foothills comprise the Foothill Sonoran, the characteristic species being Digger Pine and Blue Oak, while just above these we have the Chaparral Sonoran with its characteristic chaparral species such as Buck Brush and various species of Manzanita. The Valley Sonoran, Foothill Sonoran and Chaparral Sonoran taken together comprise the Upper Sonoran, one of the two subdivisions of the Sonoran Zone.

The Transition Zone has an average altitude of 2000 to 5000 feet with a mean annual temperature of 55 to 60 degrees and an average rainfall of 25 to 35 inches. This zone in the Sierra Nevada corresponds to the main timber belt and is composed of five dominant forest tree species. It is usually well defined on its lower borders where it joins the Sonoran Zone; sometimes indeed the tension line between the zones is so sharp as to be almost dramatic. It is along or near this tension line that the California Black Oak is common, a species very frequently associated with the most wide-spread species of the Transition, the Western Yellow Pine. The other widely-spread species are Incense Cedar, White Fir, Sugar Pine and Big Tree. The Transition Zone also extends southward over the slopes of the high mountains of coastal Southern California at average altitudes of 4000 to 7000 feet and carries all the tree species just noted except the Big Tree.

The Redwood belt also falls within the Transition Zone and is known as the Redwood Transition, for although the Redwood is found from sea-level to 3000 feet and thus does not correspond at all in altitude to the Sierran Transition, nevertheless, the factors of rainfall, humidity and temperature dominate to such a degree that the factor of altitude is relegated to an inferior influence. The associated species are Coast Hemlock, Tideland Spruce, Lowland Fir, Douglas Fir, Tan Oak and Madroño.

The Canadian Zone comes next above the Transition Zone. In the Sierra Nevada it has an average altitude of 5000 to 7000 feet, a mean annual temperature of 50 to 55 degrees and a rainfall of 40 to 50 inches. The characteristic forest species of the Sierra Nevada Canadian are the Red Fir, Silver Pine, Tamrac Pine, and Jeffrey Pine. This zone also appears on the more elevated summits of the Coast Ranges and of the high mountains of Southern California.

The Hudsonian Zone in the Sierra Nevada has an average altitude of 7000 to 9000 feet with a mean annual temperature of 40 to 50 degrees and a rainfall of 50 to 55 inches. Within this zone grow the timber line trees, the White-bark Pine, Mountain Hemlock, and Sierra Juniper. On some peaks Tamrac Pine is a timber line species. This zone recurs only in a limited degree on the very high summits of the Siskiyou and Trinity mountains in the North Coast Ranges.

The Boreal Zone in the Sierra Nevada has an average altitude of 9000 to 14,500 feet, with a mean annual temperature of 40 to 50 degrees and a rainfall of 60 to 70 inches. It supports no trees but only herbs, woody-based plants or small depressed shrubs.

These zones must not be thought of as sharp and definite bands like lines of latitude and longitude marked off on a map. They often thrust out irregular belts or tongues below or above their ideal levels. Small areas of certain species, known as "islands," and surrounded by species which belong to a different zone, may in this manner be entirely separated physically, though not physiologically, from the main body of their zone. This condition results from the operation of various local factors and certain variations of the main factors. The chief influences causing irregularity in the life zones are as follows: (1) North and east slopes are moister than south or west slopes and forest species of a given zone tend to run lower, sometimes very much lower on the former slopes. (2) South and west slopes are often excessively dry or arid and the zone is pushed higher, often very much higher at such points. (3) The mean temperature rises as we proceed southward so that a given zone lies at a higher altitude in the south than in the north. (4) The distribution of rainfall over long periods rather than its amount may often localize certain species in small "islands" beyond the limits of the main zone. (5) Series of springs or seepage slopes may localize or separate "islands" from the main zone.

In spite of these local departures the concept of Life Zones as applied to California results in useful categories. The zones are for the most part strikingly evident to the field student. The change from the Upper Sonoran Zone to the Transition Zone is often so sharply defined that it can be localized within a few yards or even within a few feet. Similarly, the limits between the Hudsonian and Boreal zones are well defined. The Canadian Zone, however, is with us not so readily defined or its limits, if they exist, are as yet not well understood.

Forest Provinces

On account of the peculiar topography of the State, the height, direction and ramification of the mountain ranges, the varying rainfall due to altitude or distance from the ocean, and the variations in temperature, it is also convenient to divide the state into six forest or silvical areas. These merge more or less on their boundaries but they are on the whole physiographically natural divisions and extremely convenient for citation. They are as follows: Sierra Nevada, North Coast Ranges, South Coast Ranges, Sacramento and San Joaquin Valleys, cismontane Southern California and the desert region (Mohave and Colorado deserts).

1. Sierra Nevada

The Sierra Nevada is a lofty and unbroken mountain range, 500 miles in length and 6000 to 15,000 feet in height, its western base rising from a plain that is only about 500 feet above sea-level. The eastern slope is very abrupt and about 5 to 15 miles wide. The western slope, which bears the main timber belt, is about 40 miles wide and is comparatively gradual. The foothills, 500 to 3000 feet, are barren or support only a scattered but characteristic growth of Digger Pine, Blue Oak and Interior Live Oak. The main timber belt begins at 2000 feet in the north and 5000 feet in the south and consists in its virgin condition of a most magnificent stand of coniferae, the four most abundant species being Yellow Pine, Incense Cedar, White Fir and Sugar Pine, the preponderance of individuals in the order named. Black Oak is usually found with Yellow Pine in the lower part of the Yellow Pine belt or just below it. Black Cottonwood usually occurs in cañon bottoms. Big Tree is also found in this belt and is often the dominant species in restricted areas, although always associated with the four conifers just named. The upper portion of the main timber belt is characterized by the presence of the Silver Pine, Red Fir and Tamrac Pine. Above the main timber belt occur the timber line trees, most of them with conical trunks excessively thickened at base, short branches, and irregular or broken tops. These include the Whitebark Pine, Foxtail Pine, Mountain Hemlock and Sierra Juniper.

2. North Coast Ranges

The North Coast Ranges of California comprise the ranges from San Francisco Bay north to the Oregon line. The main feature of this forest province is the remarkable development of the Redwood Belt, from the lumbermen's standpoint the densest body of timber in the world. The Redwood is the dominant tree in the Redwood Belt, the subordinate species being Douglas Fir, Lowland Fir, Sitka Spruce, Coast Hemlock, Canoe Cedar, Big-leaf Maple and Oregon Ash. All the subordinate species are derived from the north, that is, they have their greatest development in the great forests of the Pacific Northwest. East of the Redwood belt is the Tan Oak belt, consisting of the Tan Oak, Black Oak, Maul Oak, Oregon Oak, Madroño and Douglas Fir. Tan Oak and Madroño also occur in association with the Redwood. The high inner North Coast Ranges with their forests of Yellow Pine, Sugar Pine, Incense Cedar, White Fir, and Sierra Juniper simulate on a small scale the timber belt of the Sierra Nevada. The foothills are usually thinly timbered with Blue Oak and Interior Live Oak.

The beautiful and charming Coast Range valleys of Napa, Sonoma, Santa Rosa, Berryessa, Scott, Ukiah and Little Lake have characteristic silvical features in the scattered groves of Valley Oak and of Live Oak which adorn their plainlike floors.

3. South Coast Ranges

The South Coast Ranges, from San Francisco Bay south to the north boundary of Santa Barbara County, is a land almost destitute of real forest save for the narrow tongue of the Redwood belt protruding south along the coast in the Santa Cruz and Santa Lucia mountains and saving also scattered patches of Yellow Pine on the summits of the Santa Lucia, Gabilan, and Mt. Hamilton ranges. Otherwise the tree growth on the rolling hills and valley levels consists of thin or scattered groves of Coast Live Oak, the most abundant and widely distributed tree in this region, Valley Oak and Blue Oak, or occasionally Digger Pine. Leagues and leagues of hills in this area are quite treeless, since the province as a whole is naturally semi-arid.

Silvically this province is remarkable for the number of species inhabiting the coast line which have a local or exceedingly restricted distribution. The singular coast species are the Monterey Pine, Bishop Pine, Santa Lucia Fir, Monterey Cypress and Gowen Cypress.

4. Sacramento and San Joaquin Valleys

The Sacramento and San Joaquin valleys, collectively termed the Great Valley, present for the most part vast areas of treeless plains. The region is naturally semi-arid in the relation of climate and native vegetation. The arboreal growth is wholly confined to river banks or bottoms, river benches or moist deltas or alluvial lands. Valley Oak is everywhere the most characteristic growth on the fertile loams (Fig. 47). It is well nigh the only widely dispersed tree in the Great Valley save for the fringe of willows, Oregon Ash, White Alder and Fremont Cottonwood along stream banks or in river bottoms.

5. Cismontane Southern California

Cismontane or coastal Southern California is separated by a series of high mountain chains, having a northwesterly and southerly trend, from the deserts. The term is here meant to apply to the whole westerly slope of this dividing sierra and thence west to the ocean. On these mountain ranges, from 5000 to 11,000 feet, is found a forest flora which, as to most of its species, is an extension southward of the Sierra Nevada forest at an altitude sufficient to insure favoring climatic conditions. The more important species and varieties on the summits and westerly slopes are Yellow Pine, Jeffrey Pine, Sugar Pine, White Fir and Incense Cedar. The foothills and mesas are characterized by Coast Live Oak and Mesa Oak, while Western Plane is a feature of the washes or beds of flood streams. In addition there are the peculiar coastline types, Island Oak, Catalina Ironwood and Torrey Pine.

6. Mohave and Colorado Deserts

The trees most characteristic of the arid province are desert types such as One-leaf Piñon (Fig. 23), California Fan Palm, Joshua Tree, Mohave Yucca, Smoke Bush, Honey Mesquite (Fig. 120), Screw-bean Mesquite (Fig. 121), Palo Verde (Fig. 122), Desert Ironwood and Desert Willow.

Southern California is weak in its forest development but is remarkably rich in species of trees. This is because it has high mountain ranges situated not far from the ocean and rising out of a low-lying desert country. There are thus brought very near together three distinct silvas, the desert silva, the high montane silva and the peculiar coast silva.

Tree "Islands"

The coast of California is forestrally and geographically interesting for the number of local species which grow along it. The range of all of them is discontinuous, and nearly all of the mainland species recur on one or more of the Santa Barbara Islands or on islands off the Lower California coast southward. Most of these species are conifers, most are strictly littoral and most of them occur in few and widely separated localities. Such localities when very circumscribed or well-defined geographically, or by their plant composition set off rather sharply from the surrounding flora, are here called "islands." The best and most striking example of an arboreal island is that at Monterey where the Monterey Pine, a local species and the dominant tree on the Monterey Peninsula, is confined to a very limited area about five miles square. With it there are four other conifers, Bishop Pine, Knob-cone Pine, Monterey Cypress and Gowen Cypress. Bishop Pine occurs sparingly at Monterey; it recurs on the north coast and at San Luis Obispo on the south coast. It is also found on Cedros Island and at one station of small area on the Lower California mainland. Knob-cone Pine is local in small quantity on the Monterey Peninsula. It is widely distributed through the Coast Ranges and Sierra Nevada but the localities are few and widely separated, and with few individuals in a locality except in the far northern part of its range. Monterey Cypress is a strictly local species not occurring elsewhere, although the Guadalupe Cypress of Guadalupe Island is very closely allied. Gowen Cypress occurs at Monterey as a dwarf and not elsewhere. This formation at Monterey is a rather remarkable island since the five conifers are confined to a small littoral area and with one exception are not found elsewhere in the immediate region.

Another "island" of Monterey Pine occurs at Pescadero on the Santa Cruz coast sixty miles northerly from Monterey. A third "island" is at San Simeon on the San Luis Obispo coast eighty miles southward. Monterey Pine does not occur elsewhere on the California mainland but is found on Santa Cruz, Santa Rosa and Guadalupe islands.

An island of Bishop Pine is found on the Point Reyes Peninsula; it is about ten miles long and one mile wide. This species occurs again in a narrow strip on the Sonoma and Mendocino coasts, being associated in Mendocino with the local Mendocino Cypress. The most northerly island of Bishop Pine is found on Luffenholz Creek near Trinidad on the Humboldt coast.

Torrey Pine is restricted to a small area about eight miles long and one and one-half miles wide on the San Diego coast at Del Mar. It occurs not elsewhere save on Santa Barbara Island. Island Oak is strictly insular, being found on Santa Catalina and Guadalupe islands. Catalina Ironwood is confined to four of the Santa Barbara Islands, Santa Catalina, San Clemente, Santa Rosa and Santa Cruz. In this connection may be noted the peculiar Santa Lucia Fir which inhabits only the Santa Lucia Mountains where it is known at about ten stations.

These peculiar local species are all littoral and all confined to a few localities of limited extent. Their present representation is very meagre in individuals. They are not increasing their area but the climatic conditions of their local habitats enable them to persist. It may certainly be assumed that they once had a more extensive distribution than at present and that geological and climatic changes have narrowed them to their present limits.

At the end of the Pliocene period there was inaugurated a tremendous series of earth movements on the California coast. Geologists are by no means agreed as to the period and duration of these oscillations but in the Tertiary and Quaternary there was at intervals land connection between the present mainland and the Santa Barbara Islands. A moister climate in the Pliocene or Pleistocene periods would permit the existence of a great forest along the California coast and its extension southwards over a large land area which now rests beneath the Pacific Ocean save for the emersed peaks of the Santa Barbara Islands. Subsidence of the mountainous South Coast Range area left only vestiges of this forest on the emersed peaks or islands. Between these islands the tides flowed through the waterways (Pacheco Pass, Panoche Pass, Warthan Pass, etc.), connecting the ocean and the inland sea of the Great Valley. The final uplift of the Coast Ranges, with the species following the receding shore downwards, accompanied by changes and diversification in climatic conditions would account for the persistence and isolation of the present tree "islands" of Monterey Pine, Monterey Cypress and other species along the California coast line. Subsidence and uplift would also explain the presence of species on some of the Santa Barbara Islands and not on others by reason of the differences of altitude among the islands.

The tree "islands" along the coast are, then, here taken to be remnants of a great Pleistocene forest. In support of such a proposition is may be indicated that the species under consideration are of a few stations with few individuals, that they are living naturally within very narrow topographic and climatic limits, that they are barely holding their own in their present habitats and that evidence is at hand that the term of life of two of them, Monterey Pine and Monterey Cypress, becomes much abbreviated in the dry California interior even when living under horticultural conditions. There is also geological evidence that the former species had at one time a greater range than at present, since fossil cones have been found at Mussel Rock near San Francisco and at Preston Point on the Marin coast.

Big Tree may in a sense be said to form tree "islands" in the northerly parts of its range. Such groves as North, Calaveras, Stanislaus, Tuolumne, Merced, Mariposa, Fresno and McKinley groves are isolated remnants where the favoring physical conditions made the persistence of the species locally possible. In the southern part of its range the localities are more numerous and less sharply defined.

The "Klamath Mountains"

The "Klamath Mountains" is a designation used by Diller* for a high-montane area in northwestern California and southwestern Oregon which includes a number of mountain chains well known under the following names: Siskiyou, Marble, Scott, Salmon, Bully Choop and Yollo Bolly. This area has as its southwestern boundary the Trinity and South Fork Trinity rivers; it extends north to Rogue River, Oregon, and east to Yreka, Sisson, Redding and the upper limits of the foothills on the west side of the upper Sacramento Valley. This area is described as independent, topographically and geologically, of the adjacent Coast Ranges, both in California and Oregon. The rocks are older and harder than those of the Coast Ranges and similar to those of the Sierra Nevada. The periods of uplift and subsidence as understood by Diller are described in the paper referred to above and are given in detail from the close of the Eocene down to the present epoch. During the Miocene and certainly in the Cretaceous the Coast Ranges were submerged and the "Klamath Mountains" rose out of the sea or were bordered by its estuaries. The downward movement of 1500 feet of the whole coast of northern California and southern Oregon as late as the Pleistocene is, perhaps, the oscillation of greatest interest in connection with the phenomena of plant distribution in the area under consideration.

There are not as yet sufficient data available to correlate historically the geology of the region with the plant distribution. While climate must have been of first importance in determining the character of the vegetation and its distribution, of course climatic factors might have moved closely along with geological changes. In any event the geological history and limits of the "Klamath Mountains" and the main features of the local plant distribution when brought into one view give rise at once to many interesting suggestions. The area is noteworthy in particular for the number of species which are either peculiar to the region or do not extend into the contiguous Coast Ranges. These species as a whole if plotted on a map would duplicate very closely the area defined by Diller as the "Klamath Mountains." The most noteworthy of these species is the Weeping Spruce (Picea breweriana

*Bull. U. S. Geol. Sur., no. 196.

Wats.), strictly peculiar to this district, which is found at scattered localities throughout the central portion of the area (Siskiyous and Marble Mountain), extends north to the high mountains south of Rogue River, south to the Salmon Mountains and perhaps to the neighborhood of the Trinity Mountains (where it has been reported to exist). Deer Oak (Quercus sadleriana R. Br. Campst.) is also strictly peculiar to the "Klamath Mountains." It occurs in great abundance from Trinity Summit to the Siskiyous and northward in adjacent Oregon. Further exploration of the little-known South Fork Mountain and Yollo Bolly country may reveal it in that region. Foxtail Pine (Pinus balfouriana Jeffrey) occurs in the Scott and Yollo Bolly mountains, and is reported on Marble Mountain. It does not occur in the contiguous regions nor elsewhere save in the high southern Sierra Nevada. Aspen occurs in the Trinity Mountains but not on Mt. Shasta nor other mountains contiguous to the Klamath area. Mountain Hemlock (Tsuga mertensiana Sarg.) occurs on the high peaks of the "Klamath Mountains" but not in the Coast Ranges of California or Oregon. Shin Oak (Quercus garryana Hook. var. Breweri Jepson) is local in the "Klamath Mountains," occurring on the summits of the Scott, Marble and Siskiyou mountains. A related form (var. semota Jepson) occurs in the southern Sierra Nevada. Rhamnus occidentalis Howell occurs only in this region so far as known to the writer. Some eight or ten species of herbs occur in the "Klamath Mountains" which are not known elsewhere save locally in the southern Sierra Nevada.

A Historical Sketch of Sequoia

The genus Sequoia is represented by two living species, Sequoia gigantea Dec. (Big Tree) and Sequoia sempervirens Endl. (Redwood), confined respectively to the Sierra Nevada and Coast Ranges of California. The number of extinct species is not definitely known, although about 80 have been described. Some of these have been reduced and others are evidently referable to nearly allied extinct genera. About 35 species have been named and described from North America alone, and some 14 more have been described or listed without specific names. The material as a whole is, however, very fragmentary and there is in consequence more or less uncertainty in regard to it.

Sequoia makes its first appearance in the Lower Cretaceous where it is frequently found. It is also frequent in the strata of the Upper Cretaceous and Miocene, being widely distributed over North America, Europe and Asia. The genus is well represented in the Lower and Upper Cretaceous and Miocene of Spitzbergen and especially of Greenland where it had the company of other conifers and of broad-leaved trees. Between Greenland with its fossil species and California with its living species many intermediate fossil stations have been discovered, as at the mouth of the Mackenzie River, in Alaska, Montana, Oregon, and California. Five of the 35 extinct species have thus far been revealed by exploration of the rocks in California.

Sequoia langsdorfii Heer has been found at Hyampum in Miocene beds; it also occurs in the Miocene of the John Day Basin in Oregon and in Yellowstone Park. This species had a wide distribution both in space and time, extending from western Europe to Manchuria, Greenland, and North America and occurring from the Cretaceous to the Pliocene, being perhaps most abundant in the Miocene. Of all extinct species it is the one most nearly allied to Sequoia sempervirens. The second species is Sequoia angustifolia Lesqx. Material from Hyampum, Hay Fork, and Corral Hollow (San Joaquin County) has been referred to this species. Stations have also been determined in the John Day Basin in Oregon and in Idaho. The original material came from Elko, Nevada.

A third species is Sequoia fairbanksii Fontaine, known only from the type locality, Slate Springs, Sierra Nevada. The fourth is Sequoia penhallowii Jeffrey, from Blue Gap in the Sierra Nevada. The fifth is Sequoia reichenbachii Heer, which occurs at Horsetown in Shasta County; it had, at least in successive epochs, a wide distribution over North America, and extended into the arctic regions.

These statements here given regarding fossil species of Sequoia are as published by paleontologists, but the number of fossil species as listed needs, it would seem, considerable qualifying. It is more or less the custom of paleontologists to publish material as new when found in a new horizon or even in a new locality. Apparently many species have also been published without weighing the range of variation that exists in the organs of our living species. In view of all this it is probable that the species of Sequoia, when carefully monographed, will be reduced to seven or eight, or perhaps a smaller number.

Sequoia sempervirens has come down from very ancient times geologically without marked change. As Krausel has shown, this tree is the characteristic species of the brown coal in Germany which is of Tertiary age. It is probable that the history of Sequoia gigantea is equally impressive both in time and space. At any rate these two living species are undoubtedly the direct descendants and remnants of a great group of Miocene taxodiaceous species. The most important factors which caused the geographic segregation of these two living species on either side of the great central basin of California must have been climatic. They are even now living in essentially different climates. Sequoia gigantea inhabits a region of long mostly rainless summers with continual warm sunshine, of cold winters with moderate rainfall, with snow several feet deep and with a temperature of high daily and annual range, often in winter reaching zero or below. Sequoia sempervirens inhabits a region of high winter rainfall, nearly rainless but exceedingly foggy summers, and moderate temperature with slight daily and annual range.

Jeffrey has clearly shown that resin ducts in Sequoia gigantea are found only in the first annual ring of wood of vigorous branches of adult trees, in leaf traces of vigorous leaves of adult trees, and in the cone and its peduncle. In Sequoia sempervirens resin ducts are practically absent. The occurrence of resin ducts in organs ancestrally old and their absence from the juvenile or more recently developed parts suggests that resin ducts are passing out—that they are vestigial structures. In both species, it should be noted, resin may sometimes flow freely from wounds. Such wounds on the trunks of Redwood trees may sometimes be seen in the Mendocino and Humboldt woods, and, so behaving, indicate a reversion to ancestral conditions precipitated by mutilation.

The Miocene forest of the Sierra Nevada doubtless contained a greater area of Sequoia than now. Sequoia species may possibly have been the dominant trees in that forest, just as Sequoia sempervirens is today the dominant tree in the Redwood belt. Still it must be pointed out that on the basis of the comparative morphology of the reproductive parts, all the evidence points to the view that the genera of Pinaceae are ancestrally older than Taxodiaceae which include the Sequoias, and that the representation of Pines, Spruces, Hemlocks and Firs in the Sierra Nevada Miocene forest was probably strong in individuals as well as in species.

The Fire-type Forest of the Sierra Nevada

For unnumbered centuries the mountain and valley country of California has been subject to the influence of grass, brush and forest fires. These fires in aboriginal times were periodically set by the native tribes for the purpose of keeping open their trading, hunting or harvesting trails, for driving game, or in the practice of a rude agriculture. Wild fires still occur irregularly or sporadically, being set by the careless or the ignorant, by the criminally minded, or by lightning. But even if there be at some future time full control of wild fires, the lower as well as the higher country will show for centuries the effects of long-continued fire ravage. Both the forest as a whole and the individual trees which compose it have in certain features a distinct relation to centuries-old fire conditions.

The Sierra Nevada forest, as the white man found it, was clearly the result of periodic or irregular firing continued over many thousands of years. Over large and small areas of this region wild fires have run in irregular succession. An incredibly large number of fires originate from lightning, especially in the high mountains. In 1917 at least 350 fires on National Forests in California originated from electrical storms. As a result the Sierran forest shows marked reactions to millennial fire conditions. Three observations illustrate this statement. First of all, the finest stands in this belt occur where from the nature of the topography the fire ravage would be expected to be less severe; secondly, the individual trees are extremely well-spaced and commonly form a very open forest, the degree of openness often being in direct ratio to the age of the stand; thirdly, the trees as a whole, without regard to specific relationship, exhibit trunk bark of such unusual thickness and often of such non-inflammable character that it may be taken as evidence of protection to the tree if not of marked adaptation to fire ravage. Such thick barks are not of course absolute protection, but heavy layers of bark on the trunk serve to minimize or lessen the effect of fire and in some cases to prevent damage.

Six species are very prominent in the main timber belt of the Sierra Nevada. The most abundant is Yellow Pine which has bark 2 to 5 inches thick, a very important asset to the tree in resisting fire damage. The second most important species is White Fir which has in the mature tree a bark 2 to 4 inches thick. The bark of the third species, Incense Cedar, is dark red, to a degree non-inflammable and about 2 to 4 inches in thickness. The fourth species, Sugar Pine, has a bark 3 to 5 inches thick in adult trees. A fifth species, Big Tree, has a thick bark of dark-red fibres which is nearly non-inflammable. This bark attains a great thickness, becoming 6 to 15 inches thick, while bark actually 2 feet in thickness is known. Although the wood contains no resin and therefore ignites tardily and burns slowly, it is mainly to this peculiar bark structure that the Big Trees owe their protection from fire. Indeed the thickness and fire-resistant character of the protective bark of Big Tree in connection with the non-resinous wood, have been the really important factors in assuring a great longevity to this species. The trees are, to be sure, more or less fire-injured. It is seldom that one sees a Sequoia gigantea tree past maturity that has not a great fire scar or fire hollow on one side of its trunk; but the marvel is that the injury to the trees in the holacaust of fire should be so little. Still another important species is Red Fir which has a very thick trunk bark.

The persistence of the Sierra Nevada timber belt in spite of forest fires may be explained also in part by several other considerations. The forest is not only a thin one over very extensive areas but it is remarkably free from undergrowth or shrubs. Yet again the main timber belt lies in a region where the annual rainfall is 30 to 50 inches and where there is more or less summer precipitation. These summer rainstorms play an important part in killing wild fires. Such climatic and forestal features are factors of prime importance in lessening fire damage. Indeed the main silvical features, that is, density, reproductive power and dominance of types, are in great part expressions of the periodic fire status.

Under the influence of repeated fires the forest was sometimes seriously thinned out (as in extensive portions of the Sierra Nevada belt), often restricted to narrower limits, and often on its lower borders went over into or was replaced by chaparral, although all chaparral does not represent one-time forest areas. Forest area gone over into chaparral indicates a change in conditions initiated by fire. Such a chaparral area is drier, the run-off is greater and the soil in consequence thinner and poorer. While the chaparral area is thus no longer forest, some of the forest species may persist in the chaparral in shrub form. These shrubs represent modifications of the types in the forest, their characteristics being for the most part the result of the change in conditions initiated by fire. Trees of the forest type are represented in the chaparral by nearly allied shrub forms as follows:

Forest Form.	Chaparral Form.
Oregon Oak (Quercus garrayana).	Brewer Oak (Var. breweri) and Kaweah Oak (Var. semota).
Interior Live Oak (Quercus wislizenii).	Scrub Live Oak (Var. frutescens).
Coast Live Oak (Quercus agrifolia).	Scrub Live Oak (Var. frutescens).
Tan Oak (Lithocarpus densiflora).	Scrub Tan Oak (Var. echinoides).
California Laurel (Umbellularia californica).	Chaparral Laurel.
Buckeye (Aesculus californica).	Scrub Buckeye.

It must be noted, however, that repeated annual fires favor in certain cases or under certain conditions the development of large individual trees. If the debris or undergrowth of one year's accumulation be burned every year, adult trees are little injured. If, however, fires be prevented until after five or ten years' accumulation of forest litter and shrubs, then the uncontrolled fire will injure more severely or even consume large trees. In proof of this note the large individual coniferous trees in the Sierra Nevada which are a direct result of repeated fires set by the native tribes and by lightning. The dense forest is destroyed or thinned out, but the solitary individual tree profits by non-crowding. Large Black Oak trees, also, often develop in chaparral under repeated firing and at the expense of a dense forest. Certain trees on the other hand are very susceptible to fire. Tamrac Pine has very thin bark and fires cause extensive damage in stands of this species.

The preservation of the Redwood belt is to be explained by somewhat different considerations. The Redwood forest is much denser than the Sierra Nevada forest and has a heavier undergrowth of shrubs and vines, but is comparatively more fire-resistant since it lies in an area of high winter rainfall (35 to 50 inches) and of summer fogs. The tops of the Redwoods mechanically collect the fog moisture and through dripping from the trees and the exclusion of sun by the fog and by the forest canopy, the floor of the forest is maintained in a comparatively moist condition through the dry season. The individual Redwood tree is fire-resistant because protected by a thick non-inflammable bark (3 to 12 inches thick) covering a non-resinous wood as in the case of the Big Tree. The Redwood forest is, therefore, more fire-resistant than the Sierra Nevada forest because of climatic factors and because this highly fire-resistant type of tree is far and away the dominant tree in the Redwood forest. The Redwood has also an added advantage in its power of vegetative reproduction or crown-sprouting.

Fire-type Pines

The cones in an ordinary type of pine like the Yellow Pine or Sugar Pine after opening and shedding their seed in the early autumn or early winter, shortly fall to the ground. The behavior of some other of our pines is in this particular very different. In the Knob-cone Pine the cones persist for very lengthened periods, 15 to 25 years or more, and they may not open during that period except through the heat of a forest fire, or occasionally and partially during successive days of great heat in summer. In case of destruction of a Knob-cone Pine forest by fire the area is thus resown with its own seed and by virtue of this peculiarity of its cones it has an advantage over other species in its struggle for possession of its favored territory. This is called a fire-type pine, which is a type remarkably adapted to fire-stricken country. The seeds are sown at the most advantageous time, they germinate a high percentage in the ashy soil, and, as a result, young stands of Knob-cone Pine are often exceedingly dense and uniform.

The seedlings which follow fire come into the reproductive stage at a very early age, often at 5 to 10 years. The species by this peculiarity is therefore adapted to a very short fire interval and is thus a true fire-type pine.

Bishop Pine is likewise a fire-type pine (Fig. 24). Its cones persist in a manner similar to those of the Knob-cone Pine but do not remain closed so long. Its young stands of saplings on fire-swept areas are very dense, pure and even. Groves of various ages, each area showing great uniformity, may be studied to advantage about Inverness on the Point Reyes peninsula.

Monterey Pine is a less striking example of a fire-type pine since its cones do not remain closed indefinitely but they do open tardily. It may be noted in this connection that Digger Pine cones open tardily, 1 to 8 months after the second autumn, and that the cones may persist as long as eight years. Beach Pine on the Mendocino "White Plains" has persistent cones which remain closed for two or three years after the second autumn. Its variety, the Lodge-pole Pine of the northwest, is much more distinctively a fire-type pine. McNab Cypress is somewhat similar to the Beach Pine, its cones often remaining closed for one or two years after the second autumn.

The number of years which the cones of fire-type pines persist on the tree is not easy to determine definitely. The figures which have been printed in the books appear to rest on the natural assumption that only one circle of cones is produced on a given shoot in a season, but such is not the case. Bishop Pine produces 1 to 5 distinct whorls of cones on the season's shoot; Knob-cone Pine produces 1 to 3 whorls and Monterey Pine 1 to 3 whorls. Even Digger Pine occasionally produces 2 whorls of cones on a season's shoot.

Tree Circles

Regeneration of the individual tree frequently occurs by crown-sprouting, that is, sprouting from the root crown—a habit characteristic of a considerable number of our species. Speaking generally conifers do not crown-sprout; the felling of a pine or fir, for example, abruptly ends the life history of the individual. One notable exception is the Redwood, which crown-sprouts very freely when cut down, fire-killed or overthrown in old age. Vigorous unmutilated trees in virgin stands may also generate sprouts. Sprouts from Redwood stumps grow into saplings and poles, compete with each other like distinct individuals, the strongest survive and we finally have a circle of trees about the parent stump. Inasmuch as the stumps are usually very large the circles about them are correspondingly large. The trees of these second-growth circles are sometimes felled, whereupon a new and still larger circle of trees will arise in a similar manner outside the first circle. Regeneration in the primitive Redwood forest has taken place, in the manner above described, for untold centuries. There are at the present time evidences of circles which began their history 500 to 1300 years ago. Probably about 80 per cent. of the adult Redwood stand consists of trees which originated from crown sprouts and not from seed. The regenerative power of the Redwood is so great that it will under natural conditions maintain possession of its area against all competition.

No other tree of the coniferous class in California crown-sprouts except the Western Yew and the California Nutmeg. The former crown-sprouts rather feebly, the latter rather vigorously. When the foliage is killed by a not too intense forest fire the California Nutmeg shows considerable vigor by sprouting very freely along the main trunk and branches.

Crown-sprouting amongst broad-leaved trees is as well-nigh universal as it is exceptional amongst cone-bearing trees. The power, to be sure, differs markedly with age and in different species. The older the tree the less likely is crown-sprouting to take place. The power also varies in the same genus, as in the case of the oaks. The power to crown-sprout declines in oaks of the white oak class with age or even disappears; in oaks of the black oak class it persists more or less into extreme age. Live Oaks, Maul Oak, and California Black Oak (species of the black oak class) crown-sprout vigorously. In some white oak species the power fails in the adult tree when the trunk is cut off close to the ground, as in the Valley Oak and Blue Oak. Increasing the height of the stump in these species, especially if the cuts be made just above the first arms of the trunk, increases correspondingly the chances of the tree's survival by sprouting.

The greatest vitality in regeneration of all members of the oak family is shown by the Tan Oak. Numerous large cone-shaped latent buds are formed at the base of the trunk in most trees. Mutilation by felling stimulates these buds into activity and many sprouts develop about the stump. The number of sprouts is very variable, commonly about 10 to 50, sometimes several hundred, and as many as 1300 have been counted by the writer. Competition for light causes, of course, a heavy mortality in these sprouts since adult circles usually contain only 3 to 8 trees. Circles of this kind, the trees almost perfectly spaced and 30 to 70 or more years old, are frequently found in the Mendocino and Humboldt woods.

Madroño sprouts from the root-crown very freely. Since the trunk in this species commonly flares at the ground so as to form a more or less table-like base, the circle of new-growth poles or trees is commonly larger, that is of greater diameter, than in other broad-leaved trees, but Madroño circles are rarely as perfect as Tan Oak circles.

Other trees which crown-sprout are the following: California Walnut, all of the willows, Catalina Ironwood, Mountain Dogwood, Oregon Ash and Blue Elder. Mature White Alder does not crown-sprout when felled or in any event probably regenerates but feebly.

Barren Foothills and Treeless Plains

Extensive barren foothills are the most characteristic feature of the South Coast Range country beyond the borders of the Redwood belt, especially towards the interior, and wide stretches of treeless plains are likewise characteristic of great areas of the Sacramento and San Joaquin valleys in their original natural condition. In both the above regions the soil conditions are highly favorable for the support of heavy forests. It is because of insufficient rainfall combined with seasonal conditions that these treeless areas are naturally unforested. The winter rainfall of 10 to 20 inches is too small to support a natural forest, except in cañons, northeast slopes or moist bottoms, especially when followed by a six months' rainless season. The mortality of seedlings under natural conditions is very high or universal in the long dry summer since they cannot get their roots far enough down to avoid desiccation and tide over the first annual drought period. For this reason small patches of wood in cañons or stands of trees on protected slopes of the interior South Coast Range country extend themselves little or not at all. In aboriginal days the annual firing of the country was a matter of great importance in the discouragement of young growth since grass fires are often hot enough to kill seedlings outright. There are also other minor factors which operate in limiting extension of wooded areas. Seed may not be distributed in favorable years, heavy rains may occur during the pollination period, frosts may and often do ruin a seed crop, squirrels and other animals destroy great quantities of seed, and heavy winter floods carry seeds to places not favorable for germination.

Arboreous vegetation suffers as well as herbaceous vegetation when the amount of rainfall is far below the normal,—when the rains cease in February or March or do not begin until December or January. Sometimes the "wet season" is nearly or quite rainless and at irregular periods two or three or four "dry years" may fall together. It is during such periods that lines of Cottonwood trees, which have extended themselves in a series of wet years out into valley washes from constant streams in the foothills, die out. It is in such years, too, that trees die about failing springs in the interior South Coast Range hills.

The observations made in reference to causes of barrenness in the South Coast Ranges apply likewise to the treeless plains of the Great Valley. While trees can be readily grown on the plains under man's care, the climatic conditions and annual fires of past times limited dense growth to the river bottoms or to moist delta lands.

The only region at lower altitudes in California which supports a dense natural forest is that of the main Redwood belt where the seasonal rainfall is about 50 inches and ocasionally rises to 122 inches in some portions of the belt. This high winter rainfall, in collusion with the summer fog and the moderate temperature of slight daily and seasonal range, furnishes the conditions under which the densest forest in California has been developed, namely, the Redwood stands of Humboldt and Del Norte counties.

Dimorphic Leaves

Leaves differing in morphological position, in shape, size and indentation of margin, or in other particulars are often found on the same tree. Differences in morphological position are of fundamental importance and are usually accompanied by differences in shape and size. Such leaves are said to be dimorphic. The true pines have dimorphic leaves. In this genus the primary leaves are thin and scale-like, the secondary leaves or needles standing in their axils and with their bases enveloped by a "boot" consisting of scale-like leaves

Dimorphic leaves are found in the Redwood. The ordinary leaves are linear, ½ to 1 inch long and 1 to 1¼ lines wide; they spread from the stem in two opposite rows, forming a flat spray (Fig. 4). In some adult trees the foliage in the very top of the crown consists exclusively of short-linear acuminate leaves 3 to 5 lines long and growing all around the stem (Fig. 5). These two foliage types do pass over into each other but the types are well marked and peculiar to different portions of the crown.

Two different shapes of leaves, perhaps well enough defined to be called dimorphic, in a very broad sense, grow on the Valley Oak: First, the ordinary foliage leaves which are pinnately cleft or lobed about to the middle, and, second, the leaves of the sterile shoots which are narrower and with deeper and broader sinuses. There is however no sharp line of demarcation between these two forms.

Of all our native trees Catalina Ironwood is most remarkable for dimorphic foliage. Some trees produce in the main, or perhaps exclusively, simple leaves and others, conversely, produce pinnately compound leaves. Particular trees may however bear quite generally both kinds of leaves.

Differences in shape, size and indentation may here be referred to briefly under the convenient heading of leaf variability. Nearly all of the oaks may be used to illustrate leaf variability. Maul Oak is almost characterized by its variability in foliage, both in size and character of margin. While the leaves in the crown of an adult Maul Oak are mainly entire, entire leaves and toothed leaves characteristically occur on the same branchlets. Extreme toothing or dentation in this species may be produced artificially. For example, a vigorous two-foot tree which was felled, shortly covered over the stump with a thrifty bush-like clump of new growth, the leaves of which were very numerous and very small (much smaller than the leaves of the original tree), very uniform in size, and very prickly. Spininess in this instance is accounted for by an excess of growth force, an explanation which covers similar cases in the Blue Oak, Live Oak and other species.

Variability of leaf form in the Blue Oak is very pronounced, either on the same tree or different trees. One tree will be characterized by small entire oblong leaves, another by large entire or sparingly toothed oblong or ovate leaves; others by oblong, elliptic, obovate or roundish leaves, entire, few- or many-toothed, or lobed (Figs. 50 and 51). Scrub Oak far surpasses Blue Oak in eccentricity of leaf outline; typically with oblong unequally toothed leaves it runs off into small entire leaves, very spiny leaves, or pinnately lobed leaves—all this accompanied by a corresponding range or fluctuation in size of the leaf. In such cases it may be difficult from a mere fragment of a specimen or even from ample material to determine the species. The collection, therefore, of a long series of specimens from a favorable individual, especially one which is crown-sprouting, is exceedingly helpful in determining the limits of leaf variability in a given species.

Food Products of the Native Trees

The subject of the food resources of the native trees has at this time mainly a historical or anthropological interest. Very few of our native species are looked upon as having any economic importance in these days of the white man. In other days—days past when the native tribes owned all California—the food products of forest and wood were of such economic value that the main existence of a tribe not infrequently depended upon the annual harvest from the trees.

The oaks in this particular were easily of first importance on account of the food value of the acorns, their heavy crop and their accessibility. Eleven species of white oaks grow within our borders and these white oaks were first in favor because they bear acorns less bitter than those of the black oaks. Of the white oaks no other species is so widely distributed over valley floors and bears acorns so abundantly as the Valley Oak. Indian village sites were usually situated in or near groves of these trees and particular groves and even particular trees were the special property of certain families or individuals. The acorns are large but long and narrow, the kernel very sweet and palatable when roasted. As was probably the practice of all native tribes the acorns were shelled, the kernels dried and the winter supply put away in storage baskets or willow cribs. The mode of preparation varied. Usually the kernels were put into a mortar, or on the surface of a flat rock and ground with a pestle into flour. The flour was immediately made into soup or into a heavy black bread for the day's consumption. Blue Oak—which belongs to the white oak class—also furnished a considerable food supply.

The Live Oaks and the California Black Oak belong to the black oak class; their acorns were similarly ground into flour but the flour was usually leeched to rid it of the bitter principle peculiar to the black oaks. Individual Live Oak trees sometimes yield sweet and palatable nuts which are used by white men in place of chestnuts. Tan Oak, like the black oaks, has a bitter acorn, but the tree occurs in such abundance in northwestern California that its acorn crop was extensively used by the lower Klamath and Eel river tribes. Even at this date Tan Oak is a food resource of Trinity River and New River Indians and white men may still see Indian women leeching flour in the sands of those rivers.

All of the pines with large nuts furnished food to the Indians, the most important species being the Digger Pine which is distributed in the foothills around the oval of the Great Valley and the One-leaf Piñon which inhabits the desert ranges, the desert slopes of the Sierra Nevada, or the desert slopes of the mountains of Southern California. The nuts of the former are larger than but not so palatable as those of the One-leaf Piñon which are still gathered in large quantities by the desert Indians and sold in the white men's markets throughout California under the name of piñon nuts. The harvesting of the nuts is a picturesque operation. The men strike the cones from the trees with poles; the women gather the cones into heaps where they are burned sufficiently to open, after which the nuts are shelled out. Other pines which may properly be called nut pines are the Parry Piñon (not so esteemed as the One-leaf Piñon), the Torrey Pine and the Big-cone Pine. All of the above pines are of the foothills or middle altitudes of the mountains.

Crops of nuts from pines at higher altitudes were also harvested. When it came autumn the tribe went on a long journey to the high mountains above their valleys where the Sugar Pines grow on the ridges, and celebrated their sylvan journeys by tree-climbing contests amongst the men. The large cones of the Sugar Pine carry a goodly supply of small but very sweet-kerneled seeds and were rather highly regarded. Yellow Pine, Silver Pine and White-bark Pine were other mountain pines whose seeds are large enough for Indian collection.

In addition to acorns and nuts a number of other trees furnished a variety of foods. The large seeds of the Buckeye were ground into flour which was washed to rid it of its powerful astringency. The berries of the Blue Elder and Madroño were used as food. A harvest, exceedingly important to the desert tribes, is found in the pods of the Honey Mesquite, Palo Verde, Screw-bean and Desert Ironwood, their importance being about in the order given.

The nut-like drupe of the California Nutmeg contains a rich oily kernel esteemed by the Indians. These were eaten, not raw but roasted, and not in quantity but a little at a time, as one might use olives as a condiment.

Local Tree Distribution and the Indian Tribes

In some cases it is evident that native species of trees do not follow the general laws of tree distribution in California, but are found in peculiar stations or habitats. These stations would not antecedently be expected and moreover are not geographically consistent among themselves.

California Walnut offers such a problem. The stations in central California at Walnut Creek, Walnut Grove, Wooden Valley and Gordon Valley are different in character, widely separated from each other, and removed by a great gap from the nearest species in Southern California. This gap represents an interval of 260 miles, a stretch of territory as suitable for the growth of the walnut as Southern California, but where no walnut species is known to exist in a natural state. The writer solves the difficulty in this way. These northern colonies perhaps originated as a mutant from seed derived from the south or perhaps the far south through the trading expeditions of the native tribes at a time that antedated by centuries the earliest local Spanish settlements. They are found in all cases about old Indian village sites and in the case of the Wooden Valley station the colony is gradually being exterminated by the advance of the primitive forest which is no longer held in check by the one-time occupants of the tribal settlement.

Very fine Buckeye trees are found on Indian shell mounds about San Francisco Bay. Undoubtedly the presence of these trees is connected with the storage of Buckeye balls and their use as food by the native tribes. To account for Buckeye trees in such habitats it is not necessary to assume anything more than accidental planting since it would be remarkable if some seeding did not take place in this way.

The long inhabitation of the country by the Indians and the peculiar local distribution of the Valley Oak in the rich valleys are in some way connected. These oak orchards, of great food importance to the native tribes, indicate plainly the influence on the trees of Indian occupancy of the country. The extent and nature of the relation of Indian tribal culture and the habit of the oaks cannot yet, if ever, be completely defined, although it is clear that the singular spacing of the trees is a result of the periodic firing of the country—an aboriginal practice of which there is ample historical evidence.

Some of the localities of One-leaf Piñon on the west slope of the Sierra Nevada are isolated, peculiar, and scarcely related to the distribution of the species as a whole. Since Piñon nuts were an important Indian food and were an article of barter on Indian trade routes far beyond the natural range of the species it may be possible to explain some outlying stations as due to Indian agency.

Effects of Earthquakes on Trees

There is current in California an old-time statement to the effect that earthquakes are sometimes sufficiently severe to snap off trunks of trees above the base. The great earthquake of 1906 offered opportunity for studies on the behavior of trees subject to seismic action. The writer was so unfortunate as to be on the opposite side of the world at that time but he returned as soon as practicable to California, reluctant to miss wholly the results of so interesting a phenomenon.

The seam of the Portola rift in Marin County gave special opportunities for observation. In a cañon running west from the Point Reyes road, a little distance westerly from Point Reyes station, the tangle of down timber was very considerable in places but always where land-slips had been started in steep cañons as a result of seismic forces. The trees here were invariably uprooted, except in a few cases where they were broken off above the roots. He saw no trees snapped off except those with rotten trunks which might as easily have come down in a windstorm of gale force.

The fault line or rift itself was 3 or 4 feet wide, and in some places, as at the Shafter Ranch near Olema, open to the depth of 4 or 5 feet. The rift here leaves the valley, strikes a hill and runs along side the hill parallel to the valley, being some 50 feet above the valley floor. This side-hill is forested with Douglas Fir, Tan Oak, Coast Live Oak and Madroño. Along the rift itself there was a considerable overthrowing of trees. The earth blocks on the opposite sides of the rift were offset about 16½ feet, that is the Point Reyes Peninsula was longitudinally displaced with reference to the mainland a little over 5 yards. Trees standing directly on the fault with their root systems spreading to opposite sides of it were, therefore, violently sheared and the Douglas Firs were overthrown in greater numbers than other trees because of their shallow rooting and absence of tap roots. It was truly dramatic to see these Douglas Fir trees extended on the ground—lying out across the valley floor at right angles to the rift, 100 to 150 feet long and 1½ to 3 feet in diameter at the base of the trunk. An overthrown Live Oak tree was 40 feet high and 2 feet in trunk diameter.

At the Skinner Ranch, a mile northerly from the Shafter Ranch, the fault line runs through the yards of the ranch buildings. Three trees of Eucalyptus globulus, standing near the edge of the rift, were moved 16½ feet without disturbing either their erect or relative position. One of them was thus brought opposite a fourth which stood on the other side of the rift, now only a few feet away.

In Mendocino County a Redwood tree, which stood directly on the fault line, had its trunk split upwards some 10 feet as a result of the strain exerted by the two halves of its root system moving in opposite directions.

It seems to be well established by credible witnesses that the crowns of trees, such as the Valley Oak, may sway very noticeably during the maximum period of oscillation at the time of an earthquake, but more serious effects have not thus far been established. Throughout the region of the Portola fault the writer saw no sound tree whose trunk could properly be said to be snapped off as a result of direct oscillation of the earth during the seismic disturbances of 1906. Only trees actually standing on the fault line were overthrown.

Common Names

Each person who uses this book should make it his prime object to learn the scientific names of the genera, family and species. This is not at all difficult, will give him accurate designations, such as are not afforded by common names, and also afford him the power of talking intelligently with foreigners who as a matter of course use the latin names. There is however so much inquiry concerning the use of certain vernacular names that it is thought proper to add here a chapter on common names.

The common names of California trees have been derived in the main from three sources: 1. From the Spanish language through the Spanish-Californians. Such names as Piñon, Madroño, Palo Verde, Encina and Roble are heritages from the Spanish occupation. Whilst their meaning and application have been more or less changed and limited during the period of transplantation from Spain, they are as employed by us useful and definite, and especially valuable on account of their historical significance and musical quality.

2. From the folk, i. e., folk-names originating in California. Names of this class form the most valuable common names because they are usually definite, pointed, and full of meaning. Here belong Sugar Pine, Silver Pine, Digger Pine, Big-cone Pine, Big Tree, Redwood, Maul Oak, Blue Oak, Valley Oak or Weeping Oak, and others.

3. From imported folk-names, i. e., English folk names of species of the eastern United States or of Europe, applied by travelers or settlers to more or less nearly related species endemic in California. To this class belong Jack Pine, Bull Pine, Yellow Pine, Pin Oak, Jack Oak, Post Oak, and many others. The use of some of these is very loose and promiscuous. Quercus garryana (Oregon Oak) is for example called Post Oak by settlers in the North Coast Ranges. Quercus douglasii (Blue Oak) is also called Post Oak by ranchmen in some parts of the Sierra foothills. Post Oak is here rejected because, first, unnecessary, and second it is the

proper common name of an eastern oak. Such instances could readily be given at greater length.

On the other hand certain importations have been retained. Yellow Pine (or more specifically Western Yellow Pine to distinguish it from the eastern Yellow Pine) is a name kept for Pinus ponderosa because so established by usage that it is not open to change. While Pinus contorta var. murrayana is not a Tamarack (which is a tree of the eastern United States belonging to the genus Larix), this common name has become so fixed by custom that it were hopeless to dislodge it and our species may properly and conveniently bear the appellation Tamrac Pine. "Tamrac" spelled thus reflects our pronunciation and joined with "Pine" gives us a combination which is distinctive and botanically accurate. In short the decisions here made have held in view the firmer lodgment of well-established common names and the crippling or rejection of those with least claims to place.

As to generic common names, such as oak, poplar, willow, walnut and maple amongst the broad-leaved trees, there is no need of confusion. There is, however, common failure by laymen to distinguish amongst genera of the coniferous trees. Nurserymen in particular mix spruces, firs and hemlocks. The Pine Family proper consists of eight genera which may be distinguished by their leaves and cones. The genus Pinus (Pine) has needle-like leaves borne 2 to 5 in a boot-like sheath, in which particular it differs from all the other genera. Abies (Fir) has narrowly linear bluntish leaves which imprint a roundish smooth scar on the branch in falling; the cones are borne erect on the limb and fall to pieces on the tree. The cones of Picea (Spruce) and Tsuga (Hemlock) fall whole; Picea has usually bristly-pointed rigid leaves which, when falling, disjoint from a spreading peg-like base which persists and roughens the branchlet; Tsuga has soft bluntish leaves which, when falling, break away from a persistent base which is ascending and somewhat blended with the stem and not spreading.

The genus Pseudotsuga is a very peculiar and definite genus; its leaves are linear and blunt, the exserted and conspicuous bracts of the cones are notched and bear a spear-like point in the notch. The cone alone, or even the bract, would distinguish these trees from any other. Yet the genus, one of comparatively modern discovery, has no distinctive common name, although various combinations of "Fir," "Spruce" and "Hemlock" with "False" or some other word have been applied to it. It is most nearly related to the Spruces, perhaps, and our common species Pseudotsuga taxifolia has long been called Douglas Spruce by botanists. Its bark, however, is that of a Fir and not a Spruce and as woodsmen recognize a tree first of all by its bark and as they call this tree "fir," the combination Douglas Fir finds a wide and easy acceptance. Moreover, this name has been adopted by the Pacific Coast Lumbermen's Association and the United States Forest Service and will undoubtedly prevail as the common name. In this connection it is interesting to indicate that the wood of this tree when manufactured into lumber is sold everywhere on the market under the name of "Oregon Pine."

Cedrus (Cedar) of the Old World is evergreen and includes the true cedars, such as the Lebanon and Deodar Cedars; it has needle-like leaves in clusters but the cones are borne erect upon the branch and fall to pieces on the tree. All the preceding genera of the Pine Family are evergreens. Larix (Larch) is a deciduous genus with needle-like leaves. Pseudolarix is an Asiatic genus.

The Cypress Family demands mention in relation to the name Cedar. That name in an accurate botanical sense belongs exclusively to trees of the genus Cedrus but in various combinations it gives us useful and widely used names for trees of other families, especially the Cypress Family. Names of this kind are now well established for such species of the Cypress Family as Libocedrus decurrens (Incense Cedar), Thuja plicata (Canoe Cedar) and Chamaecyparis lawsoniana (Port Orford Cedar), and are freely accepted by botanists.

The genus Sequoia has two species: Sequoia gigantea (Big Tree) and Sequoia sempervirens (Redwood). The common names are to be used as indicated. The name Redwood should be used only for the coast species (Sequoia sempervirens) and the name Big Tree only for the Sierra Nevada tree (Sequoia gigantea). Literally and scientifically it is true that Sequoia gigantea is as much a "redwood" as the other and it is so called locally by settlers. On the other hand Big Tree as a name might easily be, in fact has been, applied locally to Sequoia sempervirens. There are tracts of Redwood in Humboldt County where the Redwood will average as great in trunk diameter as the Big Tree in certain tracts of the Sierra. Precedence, best custom and widest usage decree, however, that Big Tree shall be restricted to the Sierra species and Redwood to the coast species. That Sequoia gigantea lumber is marketed as Redwood is of no moment in this connection. Commercial names give no certain clue to proper designation of species. "Alaska Pine" of the lumber yards is not botanically pine but Coast Hemlock and any woodsman would recognize the tree as such in the forest. "White Pine" lumber is manufactured from Yellow Pine logs, and thus the examples might be multiplied.

Species Names of Native Trees—Their Origin or Meaning

Pinus monticola Don is so called because inhabiting the mountains. In the case of Pinus albicaulis Engelm., the white stems suggested the species name, just as the pliable branchlets indicated a similar propriety in Pinus flexilis James. Pinus balfouriana Jeffrey was named for John Hutton Balfour, Professor of Botany in the University of Edinburgh, an active member of the botanical association that sent John Jeffrey to Oregon and California in 1852. The species name in Pinus aristata Engelm. refers to the bristles on the cone scales. Pinus ponderosa Dougl. was discovered in 1826 by David Douglas on the Spokane River in what is now the state of Washington; he was so impressed by the great size of this new species of pine that he called it ponderosa.

The depressed or contorted habit of the Beach Pine suggested to Loudon the species name and it became Pinus contorta. Pinus coulteri Don was first made known by the botanical collector, Thos. Coulter, who found it in 1831 in the Santa Lucia Mountains near the Mission of San Antonio. Coulter was the first botanist to cross the Colorado Desert. He met David Douglas at Monterey in 1830 and we find Douglas writing to Dr. William Hooker, Professor of Botany at Glasgow University: "And I do assure you, from my heart, it is a terrible pleasure to me thus to meet a really good man, and one with whom I can talk of plants."

Dr. C. C. Parry, botanist of the Mexican Boundary Survey of 1849, discovered in San Diego County the species which was named Pinus parryana by Dr. John Torrey. Pinus muricata Don takes its name from the prickly bosses on the cones; Pinus radiata Don derives its name from the cones borne in circles on the branches; while the name of Pinus tuberculata Gord. is to be explained by the swollen apophyses or tubercles of the cone scales.

In the case of the Coast Hemlock, short leaves one-fourth inch long are intermingled on the same branchlet with leaves three-fourths inch long; doubtless from this fact Rafinesque named it Abies heterophylla in 1832, whence the combination Tsuga heterophylla Sarg. Tsuga mertensiana Sarg. was discovered on the island of Sitka in Alaska by Dr. K. H. Mertens, a German botanist in the Russian expedition of 1826 under command of Captain Lütke. Picea sitchensis Carr. refers to Sitka, the place of its discovery. Picea engelmannii Parry is named in commemoration of Dr. Geo. Engelmann, a physician and botanist of St. Louis, who gave his hours of recreation to the study of the pines and the oaks; he traveled in California in 1880. Picea breweriana Wats. is in memory of William H. Brewer, botanist of the California Geological Survey from 1861 to 1864.

The species name in Pseudotsuga taxifolia Britt. (Douglas Fir) refers to its yew-like leaves. In Abies concolor L. & G. the leaves are essentially of one color above and below, whence doubtless the specific name. The enormous size of the trunks explains the combination Abies grandis Lindl., just as the splendid proportions of the Red Fir explain the name Abies magnifica Murr., and its wonderful pulchritude that of Abies venusta Koch.

The cypress or cedar-like trees come next. Libocedrus decurrens Torr. refers to the scale-like leaves decurrent upon and adnate to the stem. Thuya plicata Don is explained by the pairs of opposite leaves which are folded about the stem. Chamaecyparis lawsoniana Parl. is in honor of Charles Lawson, Scottish nurseryman, author of the Pinetum Brittanicum, and Lord Provost of Edinburgh (1862). The cones of Monterey Cypress are exceptionally large for a cypress, which thus explains Cupressus macrocarpa Hartw. Cupressus pygmaea Sarg. was originally named for the tiny dwarfs on the pine barrens of the Mendocino coast, before the well-developed trees were made known. Cupressus governiana Gord. was discovered at Monterey by Theodore Hartweg in 1846 and named for James R. Gowen, secretary of the London Horticultural Society of London which sent Hartweg to Mexico and California. Cupressus sargentii Jepson is in honor of Dr. Chas. S. Sargent, author of the Silva of North America and Director of the Arnold Arboretum in Massachusetts. Taxus brevifolia Nutt. received its name from its relatively short leaves.

The species name in Alnus rhombifolia Nutt. refers to the rhombic-shaped leaves. The specific part of the name in Alnus rubra was suggested to Bongard by the folk name of this tree used by the Russians at Sitka, which doubtless referred to the reddish brown wood or its red inner bark. Quercus garryana Dougl. was dedicated to Nicholas Garry of the Hudson Bay Company, from whom David Douglas received many favors while exploring Northwest America. Douglas also explored the mission coast of California, the Blue Oak, Quercus douglasii H. & A., one of his many discoveries, being in memory of him. Quercus engelmannii Greene is named for Dr. Geo. Engelmann. The species name in Quercus chrysolepsis Liebm. refers to the golden powder-like scales on the leaves and acorn-cups. Quercus tomentella Engelm. is remarkable for its tomentose leaves. Quercus palmeri Engelm. was discovered in San Diego County by the botanical explorer, Edward Palmer. Dr. A. Wislizenius of St. Louis botanized in California in 1851, Quercus wislizenii A. DC., which he collected on the American River, being named for him. Lithocarpus densiflorus Rehd. refers to the dense flowering spikes and Castanopsis chrysophylla A. DC. to the golden under side of the leaves.

Quercus kelloggii Newb. was first collected in 1846 near the Spanish town of Sonoma by Theodore Hartweg, a botanical explorer of the London Horticultural Society. It has but one folk-name, being popularly and universally known as Black Oak, because it is our only typical species of the Black Oak class. Botanists and their friends often use the common name, Kellogg Oak. Dr. Albert Kellogg was a pioneer botanist of California, one of the founders of the California Academy of Sciences, a man of singular purity of life and high character, and a devoted student of the trees. The man's nature is well revealed in his Forest Trees of California, a book to be remembered for its botanical descriptions, interspersed with spiritual interpretation of the oaks and the laurel, the pines and the madroño, and the beauty of many of its passages.

Juglans hindsii Jepson was named for Richard Brinsley Hinds, surgeon and botanist of the British exploring ship Sulphur, which ascended the Sacramento River in 1837. Hinds was the first to notice this tree and he makes record in his journal that, during the course of the expedition, "we sometimes obtained a dessert from the fruit of a juglans." Platanus racemosa Nutt. was collected by Thos. Nuttall at Santa Barbara in 1835; his specific name alludes to the distribution of the fruit balls on their axes.

Salix laevigata Bebb alludes to the shining leaves; Salix lasiandra Benth. to the hairy filaments of the stamens. Salix lasiolepis Benth. was collected by Theo. Hartweg along the Carmel and Salinas rivers; Bentham in the specific name refers to the hairy scales of the catkins. Salix sitchensis Sanson was collected at an early date near Sitka, Alaska. Populus fremontii Wats. is in honor of Captain John C. Fremont, western explorer, who lead four exploring expeditions from the Missouri River to the Rocky Mountains and California between 1842 and 1854. Populus trichocarpa T. & G. refers to the slender ovaries.

Cercocarpus ledifolius Nutt. is so-called because its leaves suggest those of the genus Ledum or Labrador Tea. Prosopis juliflora DC. produces its flowers in catkin-like clusters, whence the species name. Prosopis pubescens Benth. is so named on account of the finely hairy foliage. Cercidium torreyanum Wats. was named for Dr. John Torrey, Professor of Botany in Columbia College, long a student of western plants. In the case of Olneya tesota Gray, tesota is the native Mexican name for this species, the Desert Ironwood.

The leaves of Acer macrophyllum Pursh are the largest of American maples, wherefore the specific name. The writer has often collected leaves of this species 13 to 15 inches broad. Acer circinatum Pursh was doubtless so named in reference to the circular outline of the leaves. Negundo is the old genus name for the box elder, whence Acer negundo L. Cornus nuttallii was named by Audobon, the ornithologist, for Thos. Nuttall, one-time Professor of Botany in Harvard College, who crossed the continent with the Wyeth expedition to Oregon in 1834 and also explored the south coast of California in 1835. Richard Henry Dana, in Two Years Before the Mast, gives a lively picture of his meeting with Nuttall at San Diego.

Arbutus menziesii Pursh is named for Archibald Menzies, surgeon and botanist on the Vancouver Expedition which explored the west coast of North America from 1792 to 1794; he was the first botanical collector in California after Thadeus Haenke, and gathered specimens of many characteristic species of California. Fraxinus velutina Torr. is so called because the young leaves are velvety with a fine tomentum. The leaves in the genus Fraxinus are charteristically pinnate, but in Fraxinus anomala Wats. they are simple instead of pinnate. The willow-like leaves of Chilopsis saligna Don furnished the basis for its specific name, while the bloom on the berries makes appropriate the species name in Sambucus gluca Nutt.

Historic Trees in California

Of all the native trees, the most famous individuals are those of the Big Tree (Sequoia gigantea Dec.). The trunks of most trees in the Calaveras and Mariposa groves are marked with little marble slabs bearing the names of people of note, or of States or cities, although some of the trees are named from some quality suggested by the tree itself. The naming of the trees after favorite generals or statesmen, so characteristic of our countrymen, contributes nothing to an understanding of these primeval trees, although of some scientific advantage in that measurements have been precisely referred to definite individuals. In one hundred to three hundred years, however, most of the names will mean nothing to the curious passer-by. The trees will easily outlive the names attached to them. Indeed, some of the personal names bestowed on trees in the Calaveras Grove in the fifties or sixties of the last century are even now meaningless to the visitor except to him who knows by tradition the atmosphere of the celebrities of that earlier day.

The most widely known individual tree of Sequoia gigantea is undoubtedly the Grizzly Giant in the Mariposa Grove. Its measurements are given on page 26 above, where also may be found reference to other famous individuals. Other well-known trees in this grove are the Washington, Queen of the Forest, General Lafayette, St. Louis and Philadelphia. Very worthy of mention is the Alabama Tree, which is considered to be the most perfect and symmetrical tree in the Mariposa Grove.

In the Calaveras Grove, visited by thousands of world travelers from the time of its discovery, the trees were given individual names at early date, chiefly between 1852 and 1870. Some of the famous trees are Empire State, Keystone State, Bay State, General Jackson, Abraham Lincoln, Pride of the Forest, William Cullen Bryant and Florence Nightingale. In the South Calaveras Grove some of the most well-known trees are the Correspondent, New York and Massachusetts.

In the case of the Valley Oak (Quercus lobata Neé), individual trees of remarkable height, span of crown, or diameter of trunk are rather numerous and often have more than a local reputation. Of such that have been given individual names the most famous is the great tree on the Rancho Chico, in Butte County, which was named the Hooker Oak in honor of Sir Joseph Hooker, long time Director of the Royal Botanic Gardens at Kew, England, who in company with Asa Gray, visited the tree in 1877 as the guest of General John Bidwell. This tree is 101 feet high, with a maximum spread of crown of 147 feet. At four feet above the ground the circumference is 22 feet 3 inches. It is in every way a fine representative of its race.

The largest tree of this species, considering all dimensions, is the Henley Oak in Round Valley, Mendocino County. It is 150 feet in height and nearly 25 feet in trunk circumference at four feet above the ground. It was measured by the writer in 1897. A Valley Oak tree, long known for its size, stands on the Old Kentucky Ranch, twelve miles west of Paso Robles. It is 110 feet high, with a crown diameter of 132 feet and a trunk circumference of 28 feet 3 inches at four feet above the ground, and is called the Ayars Oak.

The Forty-niners Oak, the great oak of Big Oak Flat on the old Yosemite road, destroyed by the miners in early days, was of this species (Quercus lobata Neé) and said to be eleven feet in diameter. On the same road, about three miles beyond Groveland, is the Hangman's Oak, where, tradition has it, seven men paid the extreme penalty for robbing gold from the miner's sluice boxes in the days of placer mining.

The Maul Oak (Quercus chrysolepis Liebm.) has developed many large trees. "Old Scotty" is one of a number of fine trees of this species on the floor of Hupa Valley; it is 95 feet high, 125 feet across the crown and with a trunk diameter of 6 feet at four and one-half feet above the ground. The Boquet Cañon Oak in Santa Barbara County is 85 feet high, 125 feet crown spread, and 10 feet in trunk diameter The Tuolumne Oak is 135 feet in crown diameter and its trunk is 31½ feet in circumference at four and one-half feet above the ground; it stands on the Wiltye Ranch near Tuolumne in Tuolumne County.

The most historic individual of the Coast Live Oak (Quercus agrifola Neé) is probably the one which stands in the rear of Mission San Carlos at Monterey, and is known as the Junipero Oak. It is now reduced to little more than a stump; but a century and a half ago, on June 3, 1770, the ceremony of taking possession of California in the name of Spain was held by Father Junipero Serra beneath its spreading crown. Another equally famous individual is the Le Conte Oak, a picturesque but now aged tree which spreads above a simple stone in memory of the brothers Le Conte, one a geologist, the other a physicist, on the campus of the University of California in Berkeley.

In central San Diego County, near the San Luis Rey River, the Rincon Oak is a landmark on the western part of the great Warner Ranch. It stands about eight miles westerly from the Hot Springs and was well-known to early travelers in the region. The trunk is 25 feet 6 inches in circumference at about four feet from the ground.

The largest laurel (Umbellularia californica Nutt.) in California is the great tree on the river bench of the Russian River in northern Sonoma County. It is known as the Cloverdale Laurel. The tree is about 75 feet high and the trunk is 14 feet 7 inches in circumference at five feet from the ground. It was measured by the writer in 1902.

From the lower Mattole country in southwestern Humboldt County a wagon trail climbs to the Wilder ridge, heavily wooded with Tan Oak, Douglas Fir and Madroño, and at the end of the ridge drops again to the ford of the Mattole River. On this downward slope a mile above the French and Pixton ranch-house the road passes around a little sharply-defined shoulder, on which, sixty feet from the wagon path, stands the "Council Madroño." It is an isolated individual with clean spaces around it and as seen from some little distance up grade the tree suggests an oak on account of its broad rounded crown and exceptional size. The tree is 75 feet in height and its crown 99 feet broad in the longest direction. The trunk itself requires special description. Large trunks of Madroño (Arbutus menziesii Pursh) are usually badly hollowed or fire-burned, although appearing sound from one side; but the trunk of this tree is round and perfect, and without fire- or axe-scars or traces of disease. At its narrowest part (16 inches from the ground), it has a girth of 24 feet 1½ inches, and at the ground exceeds this measurement only by a few inches. On the south side is given off a horizontally spreading limb 4 feet in diameter. Above this limb (at 4½ feet from the ground) the circumference of the trunk is 22 feet and 1 inch. At 10 feet the trunk parts into its main branches, giving rise to a broad but very rounded and symmetrical crown. Under its spreading limbs the coast tribes met the interior tribes in former days for the discussion of inter-tribal matters and for the conclusion of treaties. Situated on a little knoll on the mountain side it commands a view of the adjacent country and has been saved from destruction or injury by fires through its local isolation in the surrounding forest. It was measured by the author in August, 1902.

The "Big Madroño of Lagunitas" stands on the south shore of the east arm of Lake Lagunitas on the north side of Mt. Tamalpais. The trunk at the smallest part (three feet above the ground), is 7 feet 4 inches in diameter; it parts into six main arms grouped in two sets of three each, and supports a crown 100 feet in longest diameter and 72 feet in height. The trunk is heavily swollen at base, the longest diameter at the ground, from tip to tip of buttresses, being 20 feet.

The occurrence of exceptionally fine individuals of the native trees prompts certain kinds of men to kill them. Tree-fallers or timber-cutters, who have worked at their trade many years, are often seized by a mastering desire, on seeing a fine tree, to cut it down. Such is the explanation of the destruction of the tallest and finest Big Tree in the Calaveras Grove immediately on its discovery and of the Forty-niner's Oak at Big Oak Flat. This chapter is not put in for diversion, but is inserted with a definite purpose. The native growth in extensive parts of California is inevitably being more or less completely exterminated and with this general slaughter historic trees are also being cut down and used as stove-wood. It is the writer's object to interest Californians in protecting individuals of special interest in all parts of California. If a tree be dedicated and marked with a stone its protection is rendered more certain. The intrinsic value of a remarkable tree reduced to cordwood is slight if the present-day cost of common labor be deducted. In those cases where legal measures cannot be employed, the California Botanical Society, with headquarters at Berkeley, will undertake in certain cases to ransome historic trees threatened by the axe.

Exploration: Far Afield and Locally

This book is one of the minor results of study and experiment upon a large amount of tree material gathered during a period of twenty years in all parts of California on many expeditions into the mountains, through the valleys, along the seashore and across the deserts. On botanical trips one may travel in part by rail, boat, stage and camp-wagon, but most of all and best of all on foot as is proper for a botanist. The most effective means of traversing wild or inaccessible regions is by means of riding animals with pack mules for carrying camp supplies and botanical equipment. Of course the botanical part of such work constitutes by no means the whole enjoyment or profit. There are all the high satisfactions which come from life in the open—the delights of the trail, the matching of one's powers or ingenuity with the elements in a storm, or with the currents and eddies of a river in flood. The botanist must lament with some others, hunters and fishers, trappers and mountaineers, that the wild places of our California are so rapidly being civilized and restricted.

However, there are still wild bits even next door to our cities and towns. All such places afford the best of laboratories (that of nature's out of doors) for the study and observation of native trees and plants, their habits, life-history, relation to seasonal conditions, reproductive power and structural characteristics. This volume is intended for use by the amateur as a means of beginning such studies. He will naturally want first to know the names of his trees, a desire which the following pages will hope to satisfy. The beginner may well use the pictures or the index of common names to determine some of the common species, but that once done he should carefully compare a known tree with its description and thereby learn how to interpret a diagnosis and how it applies to a species. The significance of descriptive terms may be further gathered by reference to the glossary. Careful work of this sort can readily be done and must be done if one would gain any real power. Illustrations are a helpful aid but cannot be relied on solely. Species which are difficult in nature cannot be made easy by pictures, but they can be made fairly clear by carefully worded diagnoses. It has not been the aim of the author to "popularize" these pages to any marked extent. A "thoroughly popularized" botany book usually contains very little botany. Such excessively diluted botanical pabulum may momentarily tickle some palates but it can afford no proper and sustaining food to a man or woman who is really and truly alive.

COAST LIVE OAK (*Quercus agrifolia* Neé) at Alameda.

SYSTEMATIC DESCRIPTIONS OF THE NATIVE TREES

Key to the Families

GYMNOSPERMS.—Naked-seeded Plants

Seeds borne naked on the surface of a scale. Stamens and ovules in catkin-like clusters (no pistils and no true flowers, that is, without an abbreviated stem bearing regular whorls of floral envelopes, stamens, and pistils).

Division I. Cone-bearing Trees. (Polycotyledons)

Cone-bearing trees or shrubs, all of ours evergreen. Trunk usually carried up through crown as a continuous axis. Leaves needle-like, narrowly linear, awl-like or scale-like. Seed-leaves 2 to 17.

Fruit a woody cone, containing several to many seeds.
- Leaves needle-like or narrowly linear; cones large or small, their scales overlapping or imbricated, each with a minute bract concealed at base on the lower side or the bract sometimes half as long as the scale or even exserted and conspicuous; seeds 2 to each scale, bearing a thin terminal wing........................Pine Family, p. 186.
- Leaves narrowly linear and 2-ranked in flat sprays, or lanceolate or awl-like and disposed all around the stem; cones small, their scales not overlapping, ending in broad flattish summits, without bracts; seeds 2 to 9 to each scale, narrowly wing-margined.............. Redwood Family, p. 196.
- Leaves minute and scale-like, thickly clothing and concealing the ultimate branchlets; cones small; scales few, overlapping, or with broad flattish summits and not overlapping, without bracts; seeds 1 to several to each scale, winged or wingless Cypress Family, p. 198.

Fruit berry-like or fleshy, one-seeded; leaves narrowly linear, in flat sprays; seed not winged..............................Yew Family, p. 202.

ANGIOSPERMS.—Flowering Plants

Seeds borne in a closed receptacle or pod. Plants with true flowers, typically with floral envelopes (calyx and corolla) and essential organs (stamens and pistils) regularly arranged on a short axis. Stamens and pistils in the same or different flowers.

Division II. Palm Trees. (Monocotyledons)

Palm or palm-like evergreen trees, the trunk simple or sparingly branched. Leaves parallel-veined, borne in a tuft at summit of stem or end of branches. Stem increasing in diameter by irregular growth, not by definite concentric layers. Flowers with the parts in 3s or 6s. Seed-leaf 1.

Leaves (in our trees) bayonet-like; flowers borne in a huge terminal panicle, not enclosed by a spathe....... .. ..Lily Family, p. 203.

Leaves ample, fan-shaped; flowers borne in large clusters enclosed by a spathe................................ ....Palm Family, p. 204.

DIVISION III. **Broad-leaved Trees.** (Dicotyledons)

Deciduous or evergreen trees, the trunk usually parting into branches, rarely persistent through crown as a continuous axis. Leaves broad, at least never needle-like in ours, netted-veined. Stem increasing in diameter by means of concentric layers. Parts of the flower in 5s or 4s. Seed-leaves 2.

A. Flowers in catkins, small and inconspicuous, never with corolla; stamens and pistils always in separate flowers; leaves alternate.

Staminate and pistillate flowers on different trees, both kinds in catkins; ovary superior; fruit a pod which splits open; seeds with a tuft of hairs; leaves simple; bark bitter; deciduous trees, mostly of stream banks or moist bottoms............................WILLOW FAMILY, p. 217.

Staminate and pistillate flowers on same tree; all deciduous trees except some of the oaks.

Leaves simple; both kinds of flowers in catkins, the pistillate catkins maturing into a woody fruit strikingly like a cone; fruit very small, 1-seeded, seed-like. ....BIRCH FAMILY, p. 205.

Leaves simple; staminate flowers in catkins; fruit a nut, either set in an acorn cup or spiny bur....................**OAK FAMILY, p. 207.**

Leaves pinnately compound; staminate flowers in catkins; fruit a nut with a husk..........................WALNUT FAMILY, p. 215.

Leaves simple, resin-dotted; both kinds of flowers in catkins; fruit a nutlet. SWEET-GALE FAMILY, p. 216.

B. Flowers not in catkins; stamens and pistils frequently in the same flower.

1. Corolla when present with distinct petals (choripetalous).

Leaves alternate.

Leaves simple, serrulate; flowers very small; fruit in ours a berry-like drupe ELM FAMILY, p. 216.

Leaves simple, large, palmately lobed, with prominent stipules; flowers minute, in ball-like clusters scattered along a pendulous thread-like axis; fruit very small, seed-like; tree of stream beds......... PLANE FAMILY, p. 220.

Leaves simple, entire, strongly aromatic; flowers in inconspicuous clusters; corolla none; stamens 9; fruit olive-like; evergreen tree......... LAUREL FAMILY, p. 220.

Leaves simple or compound; flowers (in our trees) always with both stamens and pistils; stamens 10 to many, inserted on the calyx; fruit various; wood very hard........ . ..ROSE FAMILY, p. 221.

Leaves (in our trees) compound; flowers mostly butterfly-like (i. e. with same plan as a Sweet-Pea); stamens 10; fruit a 2-valved pod (legume). PEA FAMILY, p. 224.

Leaves opposite or none.

Leaves palmately compound; flowers in a showy cylindric cluster; stamens 3 to 10; fruit a large 3-valved hanging pod; deciduous tree ...... BUCKEYE FAMILY, p. 227.

Leaves mostly simple and palmately lobed (one species pinnately compound); flowers small, mostly in hanging clusters; stamens 3 to 10; fruit a 2-winged pod (double samara).MAPLE FAMILY, p. 227

Leaves reduced to scales or prickles; spiny columnar or globose trees or shrubs; stamens numerous; flowers showy; fruit fleshy... ...... CACTUS FAMILY, p. 228.

Leaves simple; flowers small, in clusters, the clusters with showy petal-like white bracts; petals and stamens 4; fruit a tiny drupe........... DOGWOOD FAMILY, p. 229.

2. Corolla when present with united petals (sympetalous).

Leaves alternate, simple; our tree with showy clusters of white flowers, the corollas like "Lily of the Valley"; stamens 5 or 10; fruit berry-like... HEATH FAMILY, p. 229.

Leaves in ours opposite and pinnately compound; flowers minute; stamens (2) and pistils in separate flowers; fruit a winged pod (samara)........ ASH FAMILY, p. 230.

Leaves simple, narrow and long; corolla showy, 2-lipped; stamens 4; fruit a dehiscent pod; seeds winged; slender willow-like tree of the Colorado Desert................................BIGNONIA FAMILY, p. 231.

Leaves in our tree compound and opposite; flowers small, in a large flat-topped cluster; corolla regular; stamens 5; fruit a berry.................. HONEYSUCKLE FAMILY, p. 231.

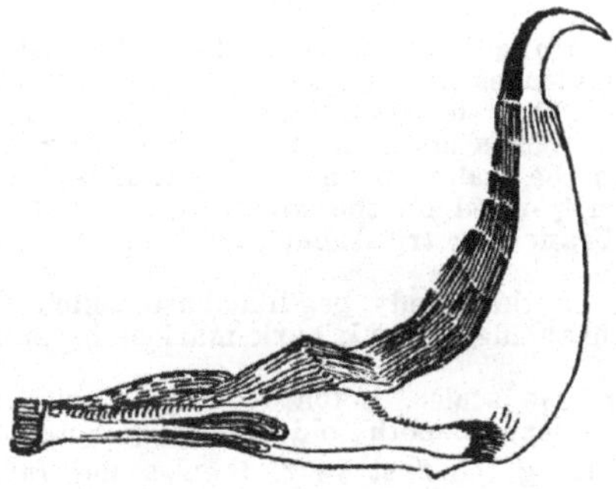

Scale of Digger Pine cone.

DIVISION I

CONIFEROUS TREES. (Polycotyledons)

Trees or shrubs, ours evergreen with linear, awl-like, or scale-like leaves. Wood resinous, the trunk usually persisting through the crown as a single axis, increasing in diameter by an annual layer of wood inside the bark. Stamens and ovules borne in clusters, generally on different branches, sometimes on different trees. Stamens generally spirally arranged in a catkin-like cluster which falls after maturity. Pollen abundant, disseminated by the wind. Ovules generally borne naked on the surface of a scale, the scales spirally arranged in a short catkin which commonly matures into a woody cone. Seed-leaves several to many, sometimes only 2. Vegetative reproduction (crown-sprouting) absent or occurring only in a few species.

PINÀCEAE. Pine Family

Trees or shrubs. Leaves narrowly linear and alternate, or with bundles of needle-like leaves in the axils of scale-like (primary) leaves. Stamens and ovules in different catkins, usually on separate branches but always on same tree. Staminate catkins with numerous spirally arranged stamens, each bearing two pollen-sacs. Ovulate catkins with spirally arranged scales, each subtended by a distinct bract; ovules naked, 2 at the base of each scale on the upper side, maturing into seeds which commonly bear a wing derived from the surface tissue of the scale. Fruit a woody cone, the scales much enlarged, the bracts elongated and surpassing the scales, shorter than the scales, or minute.

Cones pendent or spreading, falling from the tree whole, the scales persistent.
- Leaves of 2 kinds, needle-leaves in fascicles of 1 to 5 and scale-leaves; cones maturing the second year, their bracts minute... .... 1. Pinus.
- Leaves of 1 kind, linear, cones maturing in the first year, their bracts obvious
  - Bracts shorter than the scales; branchlets roughened by the persistent leaf bases
    - Leaves petioled, jointed on the woody base which is somewhat decurrent on the branchlet; trunk bark fissured or smoothish, not scaly... .... 2 Tsuga
    - Leaves sessile on the woody peg-like base which spreads at right angles to the branchlet; trunk bark marked by scars of deciduous scales .. 3. Picea
  - Bracts longer than the scales, notched at apex with a spear-like point in the notch; leaf-scars smooth; old bark very rough ....4. Pseudotsuga

Cones erect on branch, maturing the first year, their scales falling separately; leaf-scars smooth.... . ... 5 Abies

1. PÌNUS L. Pine

Trees with two sorts of leaves, the primary leaves thin and scaly or chaff-like, bearing in their axils needle-shaped leaves, in fascicles of 1 to 5, which emerge from slender buds whose scarious scales sheathe the base of the cluster. Staminate catkins spreading, crowded in a whorl towards the base of the shoot of the same spring. Ovulate catkins erect, 1 to 5 in a sub-terminal whorl. Cones maturing in the second autumn, reflexed or pendulous, their scales woody, imbricated, the exposed portion often much thickened and bearing centrally an elevated scar or prickly boss (umbo). Cotyledons 4 to 17. (Latin name of the pine.)

White Pines—Cones subterminal, the tip of the scales usually thin and unarmed; needles in 5s; wood light-colored, soft; chiefly high montane
- Cones long-stalked, very long and slender when closed; seeds shorter than their wings.
  - Needles 1 to 3¾ in. long; cones 6 to 8 in long; high ranges ...1 *P monticola*
  - Needles 2 to 3½ in. long; cones 13 to 18 in long, high ranges. 2. *P. lambertiana.*

Cones with short stalks or almost none; needles 1 to 2½ in. long; seeds longer than their wings.

Scales very thick at tip, not closely overlapping; cones subglobose, 1 to 3 in long, subalpine ....3 *P. albicaulis*

Scale-tips slightly thickened, rather closely overlapping; cones commonly long-ovate, 2 to 5 in long; desert mts chiefly ....4. *P. flexilis*

YELLOW PINES—Cones subterminal, sessile or nearly so, the scales with a thickened tip which is umbonate and armed with a prickle, needles in 5s, 3s, or 2s, wood very pitchy.

Needles in 5s

Cones oblong-ovate, 2½ to 5 in long; scales with minute prickles; needles ¾ to 1 in long; Mt. Whitney region and high North Coast Ranges .. .. 5 *P balfouriana*

Cones slender-ovate, 3 to 3½ in long; scales armed with long slender prickles, needles 1 to 1½ in long; desert ranges ....6 *P aristata*

Needles in 3s, 5 to 10 in. long, cones breaking through near base when falling, some scales remaining on branch ("broken-cone" type).

Cones ovate, 3 to 5 in long; common at middle altitudes ....7. *P ponderosa*

Cones round-oval, 5 to 8 in long; high montane .. . .Var *jeffreyi*

Needles in 2s, 1 to 2¾ in long, cones 1 to 1¾ in long, not breaking in two.

Foliage dark green; bark thick, rough; seashore . ... 8 *P contorta*

Foliage yellow green, bark thin, smooth; high montane .. ....Var. *murrayana*

NUT PINES—Cones lateral or subterminal, the scales strongly thickened at tip or prolonged into conspicuous spurs or hooks, seeds large, thick-shelled, the wing usually short or none; needles 1 to 5 in a cluster, arid areas and chiefly low altitudes

Cones very large, with highly developed spurs, breaking through near base when falling, a few lower scales persisting on the branch ("broken-cone" type)

Cones long ovate, 10 to 13 in long; needles erect, 5 to 14 in long; trunk persisting through crown as one main axis; foliage yellowish, South Coast Ranges and S Cal .... 9 *P coulteri*

Cones round-oval, 6 to 10 in long; needles drooping, 7 to 13½ in long; trunk branching into several secondary axes; foliage gray; dry interior foothills ....10. *P. sabiniana*

Cones smaller, with pyramidal tips to the scales

Needles in 5s, 8 to 12 in long; cones 4 to 5½ in. long; some basal scales persisting on the branch when falling; local on south coast. . .. 11. *P. torreyana.*

Needles not in 5s; cones subglobose.

Needles commonly in 4s, ¾ to 1⅝ in long; cones ¾ to 1½ in long; S Cal ; var parryana of.................. ..12 *P cembroides*

Needles 1 in a place, 1½ to 2 in long; cones subglobose, 2½ to 3½ in long; desert region; var. monophylla of.......12 *P .cembroides*

CLOSED-CONE PINES—Cones lateral, sessile, one-sided, opening tardily, often remaining closed for many years, their scales conspicuously swollen at tip; needles in 3s or 2s, 3 to 6 in long; lower altitudes, chiefly of coast.

Needles in 2s; cones broadly ovate, 2 to 3 in long, often developing stout spurs, seashore ....13. *P muricata*

Needles in 3s

Cones broadly ovoid, 2½ to 4½ in long, seashore . . . .14 *P radiata*

Cones oblong-ovate, 3 to 6 in long; montane..... . 15 *P. tuberculata*

1. **P. montícola** Don. SILVER PINE. Forest tree 50 to 175 ft. high, the trunk 1 to 6 ft. in diameter; bark smooth or slightly checked, whitish or reddish; needles in 5s or 4s, very slender, 1 to 3¾ in. long, sheathed at base by thinnish narrow deciduous scales, some of which are 1 in. long; cones in clusters of 1 to 7, borne near the ends of high branches on long stalks, pendulous, rather soft, 6 to 8 or rarely 10 in. long, very slender when closed and usually curved towards the tip, black-purple or green when young, 3 to 3½ in. thick near the base when open and tapering to the apex; scales thin, smooth, the umbo scar-like; seed-leaves 5 to 9.—Montane, 6000 to 9500 ft.: Sierra Nevada; Mt. Shasta to Marble Mt. and the Siskiyou Mts.; n. to B. C and nw. Mont.

2. **P. lambertiàna** Dougl. SUGAR PINE. Fig. 86. Forest tree 60 to 180 ft. high; trunk 2 to 6 ft. in diameter, its brown or reddish bark 2 to 4 in. thick, roughly fissured longitudinally with the surface breaking down into small

Fig 86. Pinus lambertiana Dougl ; cone x ⅓.

deciduous scales; needles in 5s, slender, 2 to 3½ in. long; cones pendulous, borne on stalks at the ends of the branches, commonly in the very summit of the tree, very long-oblong, 13 to 18 in. long, 4 to 6 in. thick when opened; scale tips thin, with terminal scar-like umbo; seeds 2 to 5 lines long with broad wings twice as long; seed-leaves 13 to 16.—Montane, 2500 to 7000 ft.: Sierra Nevada; Siskiyou Mts. to Yollo Bolly Mts., thence very rare and local s. to Santa Lucia Mts.; high mts. of S. Cal.; s. to L. Cal., n. to Ore. The most splendid of all pines and one of the greatest charms of the Sierran forest. Wood very valuable, white, soft and straight-grained.

3. **P. albicáulis** Engelm. WHITE-BARK PINE. Fig. 87. Dwarfish or prostrate, 2 to 6 ft. high or a tree up to 20 or 40 ft. high, often with 2 or 3 main stems from the base; needles in 5s, 1 to 2½ in. long, persisting 5 to 7 years but clothing only the tips of the slowly growing branchlets; cones ovoid or subglobose, yellowish brown, 1 to 3 in. long and nearly as thick; scales broad, at apex rounded and with a short acute umbo, not overlapping closely but their tips strongly thickened and either projecting freely or presenting very bluntish points; seed-leaves 7 to 9.—Timber line, 7500 to 12,000 ft.: Sierra Nevada from Mt. Whitney to Mt. Shasta, thence w. to the Salmon Mts.; n. to B. C. and w. Mont.

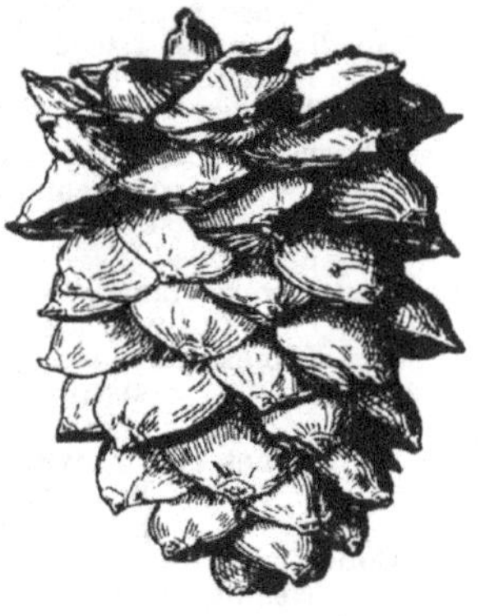

Fig 87. Pinus albicaulis Engelm.; cone x½

4. **P. fléxilis** James. LIMBER PINE. Tree 15 to 60 ft. high; needles in 5s, 1 to 2¼ in. long, often curving, densely clothing the ends of the branchlets and thus forming a sort of brush; cones buff or olive-buff, globose to long ovate, 2 to 5 in. long; scales broad with rounded slightly thickened tips and terminal scar-like umbo, overlapping rather closely and leaving only a narrow portion free on the upper side of the scale; seed-leaves 6 to 9.—Arid rocky slopes, 7500 to 11,000 ft.: San Jacinto, San Bernardino and San Gabriel mountains; e. slope Sierra Nevada in Mono and Inyo Cos.; White, Inyo and Panamint ranges; e. to Rocky Mts., n. to Alb.

5. **P. bálfouriàna** Jeffrey. FOXTAIL PINE. Tree 20 to 45 ft. high, the trunk slender-conic, 1 to 4 ft. in diameter; bark reddish brown, only slightly checked; branchlets thickly clothed with needles; needles in 5s, bright green on the upper side, glaucous on the lower, ¾ to 1 in. long, persisting 10 to 15 years; cones slender when closed, oblong-ovate when open, terra cotta in color, 2½ to 5 in. long and 1¾ to 2 in. thick; tips of the scales thickened or low-pyramidal, with shrunken scar-like umbo; seed-leaves 5.—Timber line, 6000 to 11,500 ft.: Siskiyou Mts. to Yollo Bolly Mts.; s. Sierra Nevada from head of South Fork San Joaquin River to Olancha Peak.

6. **P. aristàta** Engelm. HICKORY PINE. Tree 20 to 55 ft. high with trunk 2 to 6 ft. in diameter; leaves 1 to 1½ in. long; cones ovate, 2½ to 3½ in.

long, the scales armed with slender prickles 3 lines long; seed-leaves 6 to 7.—Desert ranges, 8000 to 11,200 ft.: Panamint, Inyo and White mountains; e. to Col. and n. Ariz.

Fig 88 Pinus ponderosa Dougl.; cone x ½.

7. **P. ponderòsa** Dougl. YELLOW PINE. Fig. 88. Forest tree 60 to 225 ft. high, the trunk 2 to 8 ft. in diameter; bark 2 to 4 in. thick, tawny or yellow-brown, divided by fissures into large plates, or sometimes closely fissured; needles in 3s, 5 to 10 in. long; cones reddish brown, commonly 3 to 5 in. long, narrowly ovate when closed, roundish ovate or oval when open, after opening breaking through near the base and falling, leaving the basal scales on the limb; scales thickened or low-pyramidal at apex and bearing an umbo which is abruptly drawn down into a stout somewhat triangular point or short prickle; seed-leaves 6 to 9.—Montane, 2500 to 6000 ft.: most abundant species in the main timber belt of the Sierra Nevada; high mts. of S. Cal.; less common in the Coast Ranges; e. to the Rocky Mts., n. to B. C., s. to Mex. Wood valuable, straight-grained, rather resinous. Passing by intergrades into the var. JÉFFREYI Vasey. Tree 60 to 170 ft. high; trunk stockier, wine-colored or yellowish; cones denser, 5 to 10 in. long, when open shaped like an old-fashioned straw hive; seed-leaves 7 to 13.—High montane, 6000 to 9000 ft.: Mt. San Jacinto and San Bernardino Mts.; Sierra Nevada; Yollo Bolly Mts. to Siskiyou Mts.

8. **P. contórta** Dougl. BEACH PINE. Mostly scrubby or dwarfish, 2 to 15 ft. high; bark dark, rough; needles in 2s, 1¼ to 2 in. long, clothing the branchlets densely, persisting 2 or 3 years; cones when closed narrowly ovate or sub-cylindric, somewhat oblique, spreading or declined when mature, 1 to 1¾ in. long, opening and releasing their seeds when fully ripe, falling after 4 or 5 years or remaining on the tree many years; tips of the scales slightly raised (low pyramidal), bearing a very slender prickle which weathers away in age; seed-leaves 4 or 5.—Sand-dunes and bluffs along the coast: Mendocino Co. to Del Norte Co.; n. to Alas. Var. BOLÁNDERI Vasey. Cane-like dwarf 2 to 5 ft. high; cones very small. — Mendocino coastal plain. In the north it passes by gradations into var. MURRAYÀNA Engelm. Tamrac Pine. Lodge-pole Pine. Fig. 89. Forest tree 50 to 125 ft. high; bark thin (¼ in. thick), smooth; needles 1 to 2¾ in. long.—Dry slopes or swampy meadows, 6000 to 11,000 ft.: San Jacinto Mts. and San Bernardino Mts.; Sierra Nevada; Mt. Shasta to the Siskiyou Mts.; n. to Alas., e. to Rocky Mts., s. to L. Cal.

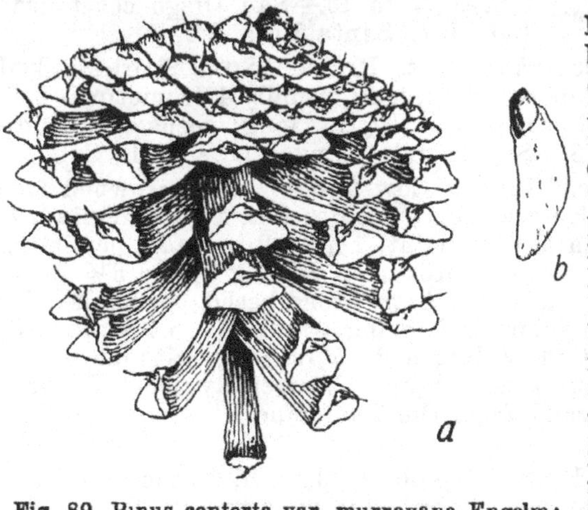

Fig. 89. Pinus contorta var. murrayana Engelm; *a*, cone; *b*, seed x 1.

9. **P. còulteri** Don. BIG-CONE PINE. Tree 40 to 90 ft. high with long lower branches; needles in 3s, erect, tipped with a short hard point, 5 to 10 (or 14) in. long; cones long-ovate, 10 to 13 in. long, 5 to 7½ in. thick, when falling breaking through near the base like the cone of the Yellow Pine; scales at tip drawn out into prominent tusk-like points or spurs which towards the base of the cone on the outer side are developed into curving talon-like appendages; seed-leaves 9 to 17.—Montane, 3000 to 6500 ft.: Mts. of S. Cal., n. to the San Carlos, Santa Lucia and Mt. Hamilton ranges; local on Mt. Diablo.

10. **P. sabiniàna** Dougl. DIGGER PINE. Fig. 90. Tree 40 to 90 ft. high, the trunk typically parting into a cluster of erect branches which form a broom-like top; needles in 3s, drooping, 7 to 13½ in. long; cones on stalks 2 to 2½ in. long, ovate, subglobose when open, 6 to 10 in. long, 5 to 7 in. thick and only slightly unsymmetrical, remaining on the tree one to seven years after releasing their seeds, when falling breaking through near the base and leaving the basal portion on the limb ("broken-cone" type); tips of the scales strongly developed into triangular hooks projecting downwards, about 1 in. long; seeds hard shelled; seed-leaves 11 to 17.—Arid foothills, forming a thin stand; Sierra Nevada foothills; inner Coast Ranges and w. to the Santa Lucia Mts.

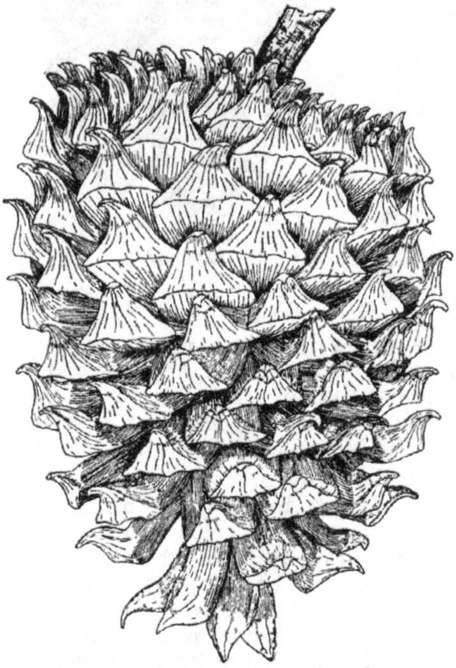

Fig. 90. Pinus sabiniana Dougl.; cone x ⅛.

11. **P. torreyàna** Parry. TORREY PINE. Widely spreading unsymmetrical tree 10 to 20 (or 40) ft. high; needles in 5s, 8 to 12 in. long; cones triangular oval, 4 to 5½ in. long, the scales at apex thickened into heavy pyramids; seed-leaves 11 to 13.—San Diego coast near Del Mar; Santa Rosa Isl.

12. **P. cembroìdes** Zucc. var. **parryàna** Voss. PARRY PIÑON. Short-trunked low tree 15 to 30 ft. high; needles ¾ to 1⅝ in. long, usually 4 (sometimes 2, 3, or 5) in a cluster; cones subglobose, 1¼ to 2¼ in. long; seeds with rudimentary wings; seed-leaves 6 to 8.—San Jacinto Range and s. to L. Cal.

Var. MÒNOPHYLLA Voss. One-leaf Piñon. Low flat- or round-crowned tree 8 to 25 (or 45) ft. high, the trunk very short; needles 1 in a place, cylindric, curving upward and ending in an abrupt point, 1¼ to 2 in. long, persisting 7 or 8 years; cones subglobose, chocolate-brown or yellow, 2½ to 3½ in. in diameter; scales thick, raised at ends into high broad-based pyramids with slightly umbilicate or flattened summits bearing a minute prickle; seeds dark brown, oblong in outline, slightly flattened, ¾ in. long, without wings; seed-leaves 7 to 10.—Common on desert ranges; extremely local (4 or 5 stations) on w. slope Sierra Nevada from the Tuolumne River to the Kern River.

13. **P. muricàta** Don. BISHOP PINE. Tree 40 to 80 ft. high, the trunk and branches with persistent circles of cones; needles in 2s, 4 to 6 in. long; cones broadly ovate, acute, 2 to 3 in. long, almost as thick, or when open more or

less globose, borne on the tree in circles of 3, 4 or 5, gradually turned downward, developed more strongly on the outside towards the base and in consequence always one-sided; scale tips rhomboidal, bearing a central prickle with a broad base, or the highly developed scales towards the base on the outside standing out as very stout straightish or upwardly curving spurs; seed-leaves 4 to 7.—Along the coast: Luffenholz Creek, Trinidad, Humboldt Co.; Mendocino and Sonoma coasts; Pt. Reyes; Coon Creek, San Luis Obispo; Mission La Purisima; L. Cal.

Fig 91 Pinus radiata Don, cone x ½.

14. **P. radiàta** Don. MONTEREY PINE. Fig. 91. Tree 30 to 115 ft. high; needles in 3s, or a few in 2s, 3 to 6 in. long; cones tan or cinnamon color, turned downward, sessile and unequally developed, broadly ovoid and bluntly pointed, or globose when open, 2½ to 4½ in. long; scales on the outer side toward the base conspicuously developed or swollen at tip into a hemispherical or pyramidal tubercle or boss and armed with a prickle which usually weathers off; seed-leaves 5 to 7.—South coast: Pt. Año Nuevo, Santa Cruz Co.; Monterey; Cambria, San Luis Obispo Co.; Santa Rosa, Santa Cruz and Guadaloupe isls.

15. **P. tuberculàta** Gord. KNOB-CONE PINE. Fig. 92. Tree 5 to 30 ft. high with thin crown and slender trunk; needles in 3s, 3 to 5 in. long; cones strongly deflexed, buff in color, narrowly ovate, oblique, acutely or bluntly pointed and somewhat curved, especially at tip, 3 to 6 in. long; scales moderately thickened at tip, except on the outside towards the base where they are raised into conspicuous rounded or pointed knobs; umbos small and contracted into slender usually deciduous prickles; seed-leaves 5 to 8.—Barren or rocky slopes at medium altitudes, the localities few and widely scattered: Santa Lucia Mts.; Santa Cruz Mts; Moraga Ridge; Mt. St. Helena; n. to Siskiyou Co., thence s. in the Sierra Nevada to Mariposa Co. The cones persist on the trunk and long slender main branches, many (15 to 25) years, forming circles from near the base to the summit; even young trees are often full of cones. The seeds are seldom liberated except when the cones are partially burned in a forest fire. It is thus interesting that a burned forest of Knob-cone Pine is promptly resown with its own seed.

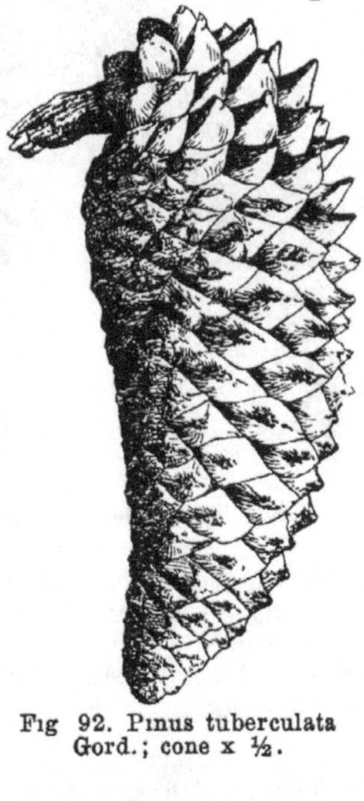

Fig 92. Pinus tuberculata Gord.; cone x ½.

2. **TSÙGA** Carr. HEMLOCK

Slender trees with nodding leading shoots. Leaves linear; resin canal 1; petioles jointed on a woody base which persists after leaf-fall as a decurrent projection roughening the branchlet. Staminate catkins consisting of a subglobose pendulous cluster of stamens on a long stipe-like peduncle arising from axillary winter buds; pollen-sacs subglobose, their cells opening transversely. Ovulate catkins erect from terminal winter buds. Cones maturing in the first autumn, solitary on ends of branchlets, pendent; scales thin, longer than the bracts. Seeds with resin vesicles on the surface; seed-leaves 3 to 6. (Tsuga, its Japanese name.)

Leaves in flat sprays; cones ½ to 1 in. long........................1. *T. heterophylla.*
Leaves spreading around stem; cones 1½ to 3 in. long..............2. *T. mertensiana.*

1. **T. héterophýlla** (Raf.) Sarg. COAST HEMLOCK. Graceful conifer, 100 to 180 ft. high, with trunk 1 to 4 ft. in diameter, the branches and branchlets slender, forming sprays which droop cascade-wise but not pendulous; branchlets finely hairy with the leaves mostly spreading in 2 ranks; leaves linear, flat, 3 to 8 lines long, ½ to 1 line wide, blunt at apex, upper side green and with a median furrow, lower side white and with a median ridge, shortly petioled; cones oblong or conical when closed, roundish when open, ½ to ¾ in. long; scales longer than broad, roundish at apex, entire; bracts about ⅙ the length of the scales, broadly triangular with truncate or obtuse summits. —Along the coast, n. Sonoma Co. to Del Norte Co.; n. to Alas. and w. Mont.

Fig. 93. Tsuga mertensiana Sarg.; fr. branchlet x ½.

2. **T. mertensiàna** (Bong.) Sarg. MOUNTAIN HEMLOCK. Fig. 93. Tree 15 to 90 ft. high, the crown with broad pyramidal base, narrowed above to a slender top; leaves standing out all around the branchlet, cylindric or somewhat flattish above, strongly ridged below, glaucous on both surfaces, bluntish at apex, ¼ to ¾ (or 1) in. long, less than 1 line wide, with a distinct but short whitish petiole; cones red-brown, rich purple when young, cylindric but tapering to base and apex, 1½ to 3 in. long, ½ to ¾ in. thick; opened cones appearing more delicate, oblong in outline or tapering from base to apex, 1 to 1¼ in. in diameter; scales thin, rounded at apex, in the opened cone spreading at right angles to the axis or even recurving, their bracts about ⅓ as long, rounded above and tipped with a short point.—Subalpine, 7000 to 11,000 ft.: Sierra Nevada from South Fork Kings River to Mt. Shasta, thence w. to the Salmon and Siskiyou mountains; n. to Alas.

3. PÍCEA Link. SPRUCE

Trees with tall tapering trunks and thin scaly bark. Leaves narrowly linear, spreading on all sides, jointed near the stem, the lower portion persistent after leaf-fall as a prominent woody base or "peg"; resin canals in ours 2. Catkins from terminal or axillary winter buds. Staminate catkins erect or nodding; anthers with nearly circular toothed crests, opening longitudinally. Ovulate catkins erect. Cones maturing in the first autumn, pendent, usually scattered over the upper half of the tree; scales thin, the bracts shorter than the scales. Seeds without resin vesicles; seed-leaves 4 to 15. (Latin name of the spruce.)

Scales of cones thin, serrulate; leaves prickly pointed or cuspidate.
- Bracts ½ to ⅔ as long as the scale; coastal ....1 *P. sitchensis.*
- Bracts ⅕ to ¼ as long as the scale; montane................ 2. *P. engelmannii.*

Scales of cone thickish, entire; leaves merely acute; high montane......3. *P. breweriana*

Fig 94 Picea sitchensis Carr ; fr. branchlet x ½.

1. **P. sitchénsis** Carr. TIDELAND SPRUCE. SITKA SPRUCE. Fig. 94. Forest tree 80 to 200 ft. high, the trunk 3 to 20 ft. in diameter and the branchlets drooping; leaves linear, ½ to 1 in. long, ⅔ to 1 line wide, whitened and flat above but with a median ridge, convex or strongly ridged below, very stiff and usually tapering to a prickly point or in top of tree less sharp or bluntly pointed; cones dull brown, long-oblong, 2 to 4 in. long and when open 1¼ to 1½ in. in diameter; scales narrow, finely and irregularly toothed, with ovate-lanceolate bracts ½ to ⅔ as long.—Along the coast, Mendocino Co. to Del Norte Co.; n. to Alas.

2. **P. engelmánnii** Engelm. ENGELMANN SPRUCE. Tree 20 to 50 (or 100) ft. high; branchlets (in ours) glabrous; leaves strongly ridged above and below, cuspidate, 8 to 11 lines long, resin-ducts often 1 or 2; cones oblong, 2 to 2½ in. long, 1½ in. in diameter when open; scales rounded or a little contracted at apex; bract ⅕ to ¼ as long as the scale.—Clark Creek, Shasta Co. (only known locality in Cal.); Ashland Butte, s. Ore.; n. to B. C., e. to Rocky Mts.

3. **P. breweriàna** Wats. WEEPING SPRUCE. Tree 20 to 95 ft. high, the thin crown ornamented with pendulous cord-like branchlets 1 to 6 ft. long; leaves ½ to 1 in. long, roundish and green below, whitish above on either side the conspicuous median ridge, obtuse; cones narrowly cylindrical, 3½ to 4½ in. long, 1¼ to 1½ in. thick when open; scales rounded at apex, very thick for a spruce and with smooth entire edges; bracts oblong, acute, ⅕ to ¼ as long as the scales; seed-leaves 6.—High cold hollows or north slopes, 6000 to 7000 ft.: Trinity Mts. to the Siskiyou Mts.; also sw. Ore.

4. PSEÙDOTSÙGA Carr.

Large trees with flat short-petioled leaves spreading around the stem or on horizontal branches often somewhat 2-ranked. Staminate catkins axillary, the pollen-sacs tipped with a spur and opening obliquely. Ovulate catkins erect, terminal or axillary. Cones pendent, maturing in the first autumn, borne all over crown; scales thin, rounded, shorter than the slender acutely 2-lobed bracts which bear a spear-like point in the notch. Seeds without resin vesicles; seed-leaves 5 to 12.—In botanical relationship it stands in an intermediate position among the Spruces, Hemlocks and Firs. Its peculiar cone bracts, signally different from those of any other conifer, and the obliquely dehiscing pollen-sacs are the chief marks of the distinctive genus Pseudotsuga. (Greek pseudos, false, and Japanese tsuga, hemlock.)

Cones 1¾ to 3½ in. long; central and n Cal . . . . 1 *P. taxifolia.*
Cones 4 to 7½ in. long, S Cal . . . . 2. *P. macrocarpa.*

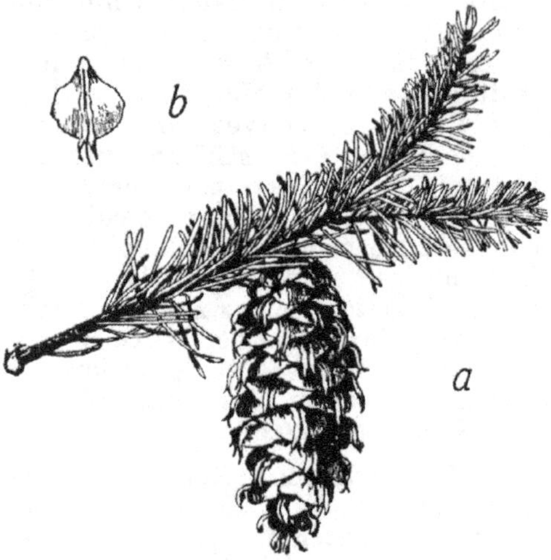

Fig. 95. Pseudotsuga taxifolia Britt; *a*, fr. branchlet; *b*, scale and bract x ½.

1. **P. taxifòlia** (Lamb.) Britt. DOUGLAS FIR. DOUGLAS SPRUCE. Fig. 95. Magnificent forest tree, 70 to 250 ft. high, the trunk 1 to 8 ft. in diameter; bark on old trees very thick, soft and putty-like, broken into broad, heavy furrows; branchlets with the leaves spreading all around the stem or on horizontal branchlets turned more or less to right and left but not in truly flat sprays; leaves linear, blunt at apex, flat with a median groove above and a ridge below, green, with two pale longitudinal bands on the under surface, very short-petioled, ½ to 1½ in. long, ½ to 1 line wide; cones cinnamon or red-brown, long-oval and more or less pointed, 1¾ to 3½ in. long, when open 1¼ to 1¾ in. thick; scales broad and rounded at apex; bracts conspicuously exserted, broadly linear and bearing in the deep notch at apex a spear-like point; seed-leaves 5 to 8.—Moist mountain slopes: Coast Ranges from Siskiyou Co. to the Santa Lucia Mts. (only near the sea southward); Sierra Nevada (3000 to 5500 ft.) from Mt. Shasta to Fresno Co.; n. to B. C. Most valuable W. Am. structural timber; marketed as Oregon Pine.

2. **P. macrocárpa** Mayr. BIG-CONE SPRUCE. Tree 30 to 60 (or 90) ft. high; leaves slightly curved; cones 4 to 7½ in. long, 2¼ to 3 in. thick when open; bracts protruding little or not at all beyond the scales, except the lowest, the tails of which are often as much as ¾ in. long; seed-leaves 6 or 7.—San Emigdio and Santa Inez mountains to the San Gabriel and San Bernardino mountains, and s. to Hot Springs Mt. and the Cuyamaca Mts., 2500 to 8000 ft.

5. ÀBIES Link. FIR

Highly symmetrical trees of lofty stature, the branches in regular whorls and ramifying laterally, forming flat sprays. Leaves linear, flat, thickened or 4-angled, whitened beneath, spreading in 2 opposite directions or even 2-ranked, or more often curving upwards, leaving a smooth circular scar when they fall; resin canals in ours 2. Catkins from axillary winter buds. Staminate catkins borne on the under side of the branches, mostly in the upper half of the tree. Ovulate catkins erect, on the upper side of the topmost spreading branches. Cones erect, maturing in the first autumn, falling to pieces on the tree; scales thin, incurved at the broadened apex; bracts often exserted. Seeds with resin vesicles; seed-leaves 4 to 10. (The Latin name.)

Leaves of lower and uppermost branches slightly different; bracts not bristle-like
- Cones 2 to 5½ in long; bracts not exserted.
  - Leaves glaucous or dull green with stomata above, flat or on cone-bearing branches keeled above, acute or rarely notched at apex, spreading in two ranks or curving upwards, with a twist in the short petiole; old bark roughly and deeply furrowed, drab or grayish; high Sierra Nevada and Coast Ranges, chiefly 3800 to 6000 ft . . . .1 *A. concolor*
  - Leaves dark lustrous green, no stomata above, white beneath, notched at apex, usually spreading in two ranks, on cone-bearing branches often blunt, curving upwards, bark white, smooth or fissured into low flat ridges; north coast only . . . . .2. *A grandis*
- Cones 4 to 8 in long, the bracts concealed or exserted, leaves thickened below and a little above so as to be subterete or somewhat 4-sided, thicker on the uppermost branches, curving upwards but not twisted, sessile, old bark deeply divided into roughly broken ridges, reddish brown, high Sierra Nevada and Coast Ranges, chiefly 6000 to 9000 ft . . . . . . .3 *A magnifica*

Leaves alike all over tree; cones with conspicuous bracts, the exserted portion long and bristle-like, bark light brown, smoothish, Santa Lucia Mts only 4 *A venusta*

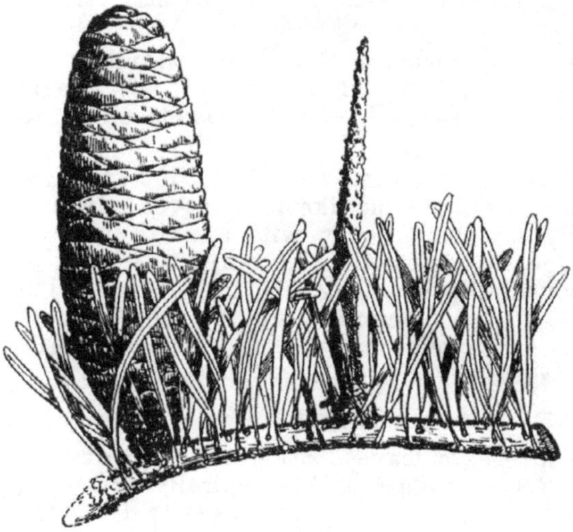
Fig 96 Abies concolor Lindl. & Gord ; fr. branchlet x ½

1. **A. cóncolor** Lindl. & Gord. WHITE FIR. Fig. 96. Forest tree 60 to 200 ft. high with a narrow crown and trunk 1 to 6 ft. in diameter; bark whitish or silvery in young trees, on old trunks gray or drab-brown, heavily fissured; leaves ½ to 2½ (commonly 1 to 2 in.) long, flat, often with a median channel on upper side (or on the uppermost branches keeled), a prominent midrib beneath with a broad depressed stomatal band on either side, acutish, obtuse or slightly notched at summit, contracted at base into a very short twisted petiole, spreading in 2 ranks or more or less erect; cones brown, oblong, rounded at summit and base, 2 to 5½ in. long, 1¼ to 1¾ in. thick; scales broad and rounded; bracts nearly ½ as long as the scales, roundish and finely toothed, often with a notch at apex and usually terminating in a short slender point; seed-leaves 6.—Mountain slopes, 3000 to 7000 (or 10,000) ft.: Sierra Nevada; high North Coast Ranges; high mts. of S. Cal.; e. to the Rocky Mts. Also, but wrongly, called Silver Fir.

2. **A. grándis** Lindl. LOWLAND FIR. Forest tree 40 to 160 ft. high, the trunk 1½ to 4 ft. in diameter; leaves flat, 1 to 2 in. long, notched at apex, dark lustrous green above and with a median channel, below with two white bands separated by a ridge; cones long-oblong in outline, 2½ to 4 in. long, 1½ to 1¾ in. thick; scales with a broad rounded summit and narrow stalk-like base, broader than long; bracts small, with a short awl-like point set on

the roundish apex, ½ as long as the scales; seed-leaves 6.—Low hills or valleys near the sea: n. Sonoma Co. to Del Norte Co.; n. to B. C.

3. **A. magnífica** Murr. RED FIR. Splendid forest tree 60 to 200 ft. high, with a trunk 1 to 5 ft. in diameter and a very narrow crown composed of numerous horizontal strata of fan-shaped sprays; bark on young trees whitish or silvery, on old trunks dark red, deeply and roughly fissured; leaves ¾ to 1½ in. long, thickened below and a little above so as to be subterete or somewhat 4-sided, sessile by a thick foot, a little contracted above the foot, acutish at apex, those on the under side of the branches spreading right and left, those in the top of the tree more thickened, erect, incurved and hiding the upper side of the branch; cones 4 to 8 in. long, 2½ to 3½ in. thick, broadly oval or oblong, the broad scales with upturned edges; bracts oblong, acute, not exserted; seed-leaves 9 to 13.—Mountain slopes and ridges, 5000 to 8000 ft.: Sierra Nevada; high North Coast Ranges from n. Lake Co. to Siskiyou Co. Wood straight- and fine-grained, heavy and very durable. Var. SHASTÉNSIS Lemm. Fig. 97. Bracts enlarged or spatulate at apex, mucronate, exserted.—Sierra Nevada and n. to Mt. Shasta and s. Ore.

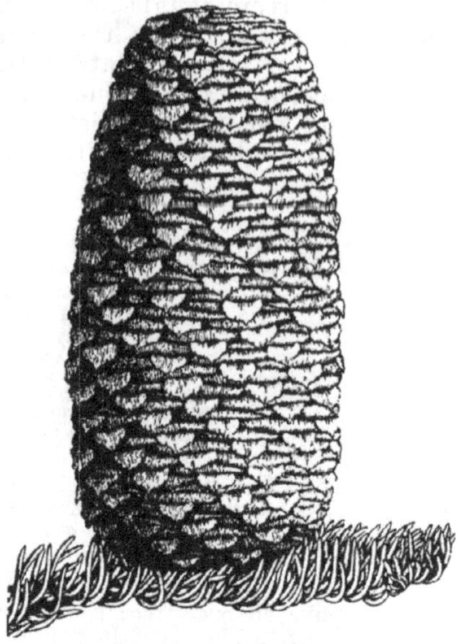

Fig 97. Abies magnifica var. shastensis Lemm ; cone x ½.

4. **A. venústa** (Dougl.) Koch. SANTA LUCIA FIR. Tree 30 to 100 ft. high, the narrow crown abruptly tapering into a steeple-like top; leaves stiff and sharp-pointed, dark green and nearly flat above, below with a white band on either side of the strong median ridge, 1¼ or mostly 1¾ to 2¼ in. long, 1 to 1½ lines wide, mostly 2-ranked; cones elliptic-oblong, 2½ to 4 in. long, 1½ to 2 in. thick, borne on peduncles ½ in. long; bracts wedge-shaped, truncate or notched at summit, the midrib prolonged into a long exserted bristle ½ to 1¾ in. long and ½ line wide; seed-leaves 7.—Santa Lucia Mts.

TAXODIÀCEAE. REDWOOD FAMILY

Trees with linear or awl-shaped alternate leaves. Staminate and ovulate catkins on the same tree. Scales of the ovulate catkins spirally arranged, more or less blended with the bract, often spreading horizontally from the axis of the cone and developed into broad flattish summits. Ovules to each scale 2 to 9. Seeds not winged or merely margined.

1. SEQUÒIA Endl. REDWOOD

Tall trees with thick red fibrous bark and linear, awl-shaped, or scale-like leaves. Cone woody, its scales divergent at right angles to the axis, widening upward and forming a broad rhomboidal wrinkled summit with a depressed center. Seeds flattened; seed-leaves 2 to 6. (The Cherokee chief, Sequoyah, who invented an alphabet for his tribe.)

Leaves awl-shaped, sessile, ascending all around stem; cones 2 to 3¾ in long; Sierra Nevada only . . . . 1 *S gigantea.*
Leaves linear, petioled, spreading in 2 ranks and forming a flat spray; cones ⅝ to 1⅛ in. long; Coast Ranges only . . . . 2. *S. sempervirens.*

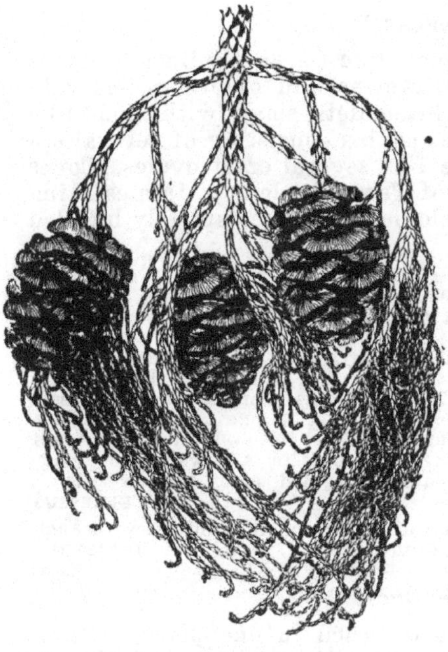

Fig 98 Sequoia gigantea Dec ; fr. branchlet x ½.

1. **S. gigantèa** (Lindl.) Dec. Big Tree. Fig. 98. Giant tree 150 to 331 ft. high, the trunk 5 to 30 ft. in diameter; bark ⅛ to 2 ft. thick, deeply furrowed; leaves awl-shaped or lanceolate, 1 to 6 lines long, adherent below to the stem which they thickly clothe; cones maturing in second autumn, red-brown, ovoid, 1¾ to 2¾ in. long, composed of 35 to 40 scales.—West slope of the Sierra Nevada, 4700 to 7500 ft., from Placer Co. to Tulare Co., occurring in scattered areas called groves or to the southward the groves or forests more extensive and less widely disconnected. The trees have been extensively lumbered and the product marketed as Redwood. The wood is similar to that of Redwood but more brittle. It is the most massive tree of the earth's silva.

2. **S. sémpervìrens** Endl. Redwood. Fig. 99. Tall forest tree 100 to 340 ft. high, the trunk 2 to 16 ft. in diameter; bark ¼ to 1 ft. thick; foliage reddish brown; leaves linear, spreading right and left so as to form flat sprays, ¼ to 1¼ (mostly ½ to ¾) in. long and 1 to 1¼ lines wide, or in the top of adult trees with short linear acuminate leaves 3 to 5 lines long, such branchlets strikingly suggestive of those of the Big Tree; cones oval, reddish brown, ⅝ to 1⅛ in. long and ⅝ to ⅞ in. thick, borne in clusters on the ends of branchlets mostly in the top of the tree, maturing in first autumn; scales 14 to 24. —Forms a narrow belt along the coast from Monterey Co. to the Oregon line, never ranging inland beyond the influence of the sea-fogs. It is the tallest tree on the American continent. No other tree has been so important to the development of civilization in California, because the wood, abundant and cheap, is highly useful for all sorts of building and industrial purposes and in manufactures and the arts.

Fig 99. Sequoia sempervirens Endl ; fr. branchlet x ½.

CUPRESSÀCEAE. Cypress Family

Trees or shrubs with opposite or whorled scale-like (or rarely linear) leaves thickly clothing the ultimate branchlets. Stamens and ovules in separate catkins. Staminate catkins terminal on the branchlets, small, with shield-like stamens bearing 2 to 6 pollen-sacs. Ovulate catkins consisting of several opposite or whorled scales which bear at base 1 to several erect ovules. Cones woody or in Juniperus fleshy, consisting of few "scales"; "scales" imbricated or shield-shaped, consisting morphologically of a completely blended scale and bract.

Fruit a woody cone; stamens and ovules on same tree
  Branchlets flattened, disposed in flat sprays, leaves opposite, in 4 rows, the successive pairs unlike; cones maturing in first autumn, seeds 2 to each scale.
    Scales of cones imbricated
      Cones pendent, scales 6, only the middle pair seed-bearing; seeds unequally 2-winged ....1 Libocedrus
      Cones reflexed, scales 8 to 12, the 2 or 3 middle pair seed-bearing; seeds equally winged . .. .2 Thuja
    Scales of cones peltate; seeds narrowly winged 3 Chamaecyparis
  Branchlets cord-like, not in flat sprays, leaves opposite, in 4 rows, alike; cones maturing in second autumn, seeds acutely margined, many to each scale .. 4 Cupressus

Fruit a berry, seeds 1 to 3 to each fruit; stamens and ovules on different trees; branchlets cord-like, leaves in whorls of 3 or opposite .. . .5 Juniperus

1. LÌBOCÈDRUS Endl.

Aromatic trees with flattened branchlets disposed in one plane. Leaves scale-like, opposite, imbricated in 4 rows, the successive pairs unlike. Staminate and ovulate catkins terminal on separate branchlets. Staminate catkins with 12 to 16 decussately opposite stamens, each with 4 to 6 pollen-sacs. Ovulate catkins consisting of 4 to 8 scales, only one pair ovule-bearing, each scale of this pair with 2 ovules at base. Cones maturing in first autumn, oblong, composed of imbricated oblong scales. Seeds unequally 2-winged; seed-leaves 2. (Greek libas, referring to the trickling of the resin, and kedros, cedar.)

1. **L. decúrrens** Torr. Incense Cedar. Fig. 100. Forest tree 50 to 125 ft. high with conical trunk 2 to 7 ft. in diameter; bark cinnamon, loose or fibrous in age; leaves minute, 1 to 3 lines long, coherent, also adherent to the stem, free only at the tips, those above and below obtuse but mi-

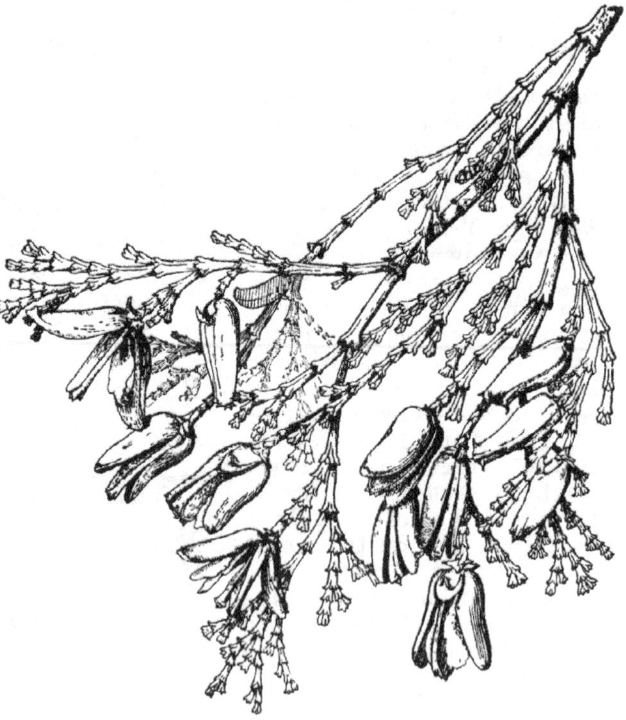

Fig 100 Libocedrus decurrens Torr.; fr. branchlet x ½.

nutely pointed and forming a pair overlapped by the keel-shaped lateral pair; cones red-brown, oblong-ovate when closed, ¾ to 1 in. long, consisting of 2 seed-bearing scales with one septal scale between them and often with 2 small scales at base; seed-bearing scales broad and flattish but not thin; all the scales with a small triangular umbo at tip; seeds 4 lines long, margined on each side from near the base to the apex by two very unequal wings; larger wing ovatish, about 6 lines long.—Montane, 2000 to 7000 ft.: common in the Sierra Nevada; higher Coast Ranges; high mts. of S. Cal.; s. to L. Cal., n. to southern Ore.

2. **THÙJA** L. Arbor-vitae

Aromatic trees with scattered branches and flattened branchlets disposed in one plane. Leaves minute, scale-like, opposite and imbricated in 4 rows, the successive pairs unlike, adnate to the stem but with free tips. Catkins terminal. Staminate catkins with 4 to 6 stamens, each with 3 or 4 anther-cells under the subpeltate crests. Ovulate catkins with 8 to 12 erect scales, each with 2 erect ovules at base. Cones small, maturing in first autumn, reflexed; scales 8 to 12, thin-leathery, the lowest and uppermost pairs sterile. Seed bordered by nearly equal lateral wings so as to be nearly round; seed-leaves 2. (Name of some tree known to the Greeks.)

1. **T. plicàta** Don. Canoe Cedar. Tree 50 to 225 ft. high with trunk 2 to 16 ft. in diameter but tapering rapidly; crown pyramidal with drooping sprays; bark cinnamon red; leaves on the margin of the flat sprays keeled or somewhat boat-shaped and acute at tip, those above and below flattish and triangular at apex; cones borne on short lateral branchlets, on opening turned downward beneath the spray, cinnamon-color, oblong in outline when closed, and ½ in. long; scales 9, the outer ones oblong or obovate, and much broader than the narrow inner ones.—Near the coast, Humboldt Co. to Del Norte Co.; n. to Mont. and Alas.

3. **CHAMAECÝPARIS** Spach

Trees or shrubs; leading shoot nodding; branchlets more or less flattened and in flat sprays; leaves scale-like, adpressed, thickly clothing the branchlets, opposite, in 4 rows, the successive pairs in ours unlike. Catkins and cones very similar to Cupressus. Stamens with usually 2 pollen-sacs. Ovules 2 to 5 at the base of each scale, the seeds winged, usually 2 (1 to 5). Cones maturing in the first autumn, their scales shield-shaped. Seed-leaves 2. (Greek chamai, on the ground, and kyparissos, cypress.)

1. **C. lawsòniàna** Parl. Port Orford Cedar. Tree 80 to 175 ft. high with drooping fern-like sprays; leaves minute, those above and below rhombodial, glandular-pitted and overlapped by the keel-shaped ones on the margin; cones globose, 3 to 4 lines long, consisting of about 7 scales; seeds 1½ to 2 lines long, narrowly wing-margined on each edge, the whole structure orbicular.—Moist hillsides or cañon bottoms: Mendocino Co. to Del Norte Co. and e. to the Sacramento River Cañon; n. to Coos Bay, Ore. In gardens called Lawson Cypress.

4. **CUPRÉSSUS** L. Cypress

Trees or shrubs. Leaves scale-like, small, appressed, closely imbricated in four ranks on the ultimate cord-like branchlets, or awl-shaped on vigorous shoots. Staminate catkins terminal on the branchlets with 3 to 5 pollen-sacs to each stamen. Ovulate catkins on short lateral branchlets, the ovules numerous, erect, in several rows at the base of the scales. Cones globose to oblong, maturing in the second autumn, the shield-shaped scales fitting closely together by their margins, not overlapping, separating at maturity, their broad summits with a central boss or short point. Seeds acutely angled or margined; seed-leaves 2 to 5. (Classical name of the Cypress.)

Umbos more or less upwardly impressed and more or less crescent-shaped.
    Bark of mature trunks dark brown, roughish and more or less fissured.
        Glands on leaves none or rare; maritime or coastal species.
            Crown with "finger-pointed" branchlets.
                Seeds brown or commonly so; umbos prominent; Monterey coast..... 1. *C. macrocarpa*
                Seeds black; umbos small; Mendocino coast. ....2. *C. pygmaea*
            Crown compact or smooth; cones with low umbos; Monterey.3. *C. goveniana.*
        Glands more or less present as a closed dorsal pit; montane species.4. *C. sargentii*
    Bark of mature trunks red, smoothish, shining; San Diego Co..........5. *C. forbesii*
Umbos more or less conical or peak-shaped; leaves with conspicuous dorsal resin pits.
    Umbos all straightish; Piute Mt, Kern Co......................6. *C. nevadensis.*
    Two uppermost umbos horn-like and incurved; northern Cal... 7. *C. macnabiana.*

1. **C. macrocárpa** Hartw. MONTEREY CYPRESS. Fig. 101. Tree 15 to 80 ft. high with broadly conical crown, the spreading branches with the terminal divisions strict and therefore "finger-pointed"; bark strands spirally twisted to the left or straight; leaves not resin-pitted; cones broadly short-cylindric or oblong, 1 to 1¾ in. long, the scales with a central curved thin-edged ridge or umbo, or the umbo sometimes subconical; seeds brown. —Headlands at mouth of Carmel River, Monterey coast.

Fig. 101. Cupressus macrocarpa Hartw.; fr. branchlet x ½.

2. **C. pygmaèa** Sarg. MENDOCINO CYPRESS. Tree 25 to 75 (or 100) ft. high, the crown with finger-pointed branchlets; or sometimes a low shrub or reduced to a dwarf cane 1 or 2 ft. high; leaves not resin-pitted; cones 5 to 9 lines long, its umbos small and low; seeds black.—Mendocino coastal plain.

3. **C. goveniàna** Gord. GOWEN CYPRESS. Shrub 1 to 5 ft. high, its outline compact or smooth; branchlets squarish; leaves without dorsal pits; cones subglabrous or oval, 6 to 8 lines long; seeds commonly black.—Monterey.

4. **C. sargéntii** Jepson. SARGENT CYPRESS. Shrub 5 to 12 ft. high; branchlets thickish, obscurely squarish; leaves or many of them with closed dorsal pits; cones 8 to 10 lines long; seeds brown.—Santa Cruz Mts. to Mt. Tamalpais, Hoods Peak Range and Mendocino Co. Var. DUTTÒNII Jepson. Symmetrical tree 40 to 72 ft. high, the branchlets more or less divaricate and the branches therefore not "finger-pointed"; bark strands spirally twisted to right or straight; leaves triangular-ovate, very short, dorsal pits none or few; cones glabrous, 8 to 12 lines long, the umbos peak-shaped; seeds brown. —Cedar Mt., Alameda Co., occupying a colony about ¾ by 1½ miles in area.

5. **C. fórbesii** Jepson. Slender tree 15 to 20 ft. high; bark very smooth, shining, red-brown or even dark cherry red; branchlets squarish; foliage bright green; dorsal pits of leaves minute or commonly wanting; cones globose, ¾ to 1⅕ in. long; seeds red-brown.—Mts. of San Diego Co.

6. **C. nevadènsis** Abrams. PIUTE CYPRESS. Tree 20 to 50 ft. high; branchlets slender, 4-angled; leaves light green with conspicuous dorsal resin pit; cones oblong to subglobose, 10 to 12 lines long; upper umbos pointed; seeds light brown.—Local on Piute Mt., Kern Co.

7. **C. macnabiàna** Murr. McNAB CYPRESS. Fig. 102 a, b. Tree 15 to 40 ft. high, or often somewhat bushy, sometimes a low shrub; foliage blue-green, pungently aromatic; branchlets very slender; leaves with a conspicuous dorsal resin pit or white gland, often slightly glaucous; cones globose to oblong, 7 to 11 lines long, reddish or grayish brown; umbos conical, the uppermost pair very prominent or horn-like and incurved; seeds brown.—Dry hills and flats from n. Napa Co. to Shasta Co., thence easterly to the n. Sierra foothills as far south as Yuba Co. It is localized in its occurrence and everywhere forms restricted groves. Var. BÀKERI Jepson. Modoc Cypress. Fig. 102c. Cones silvery or glaucous, 5 to 6 lines long; umbos short-conical.—Modoc lava beds.

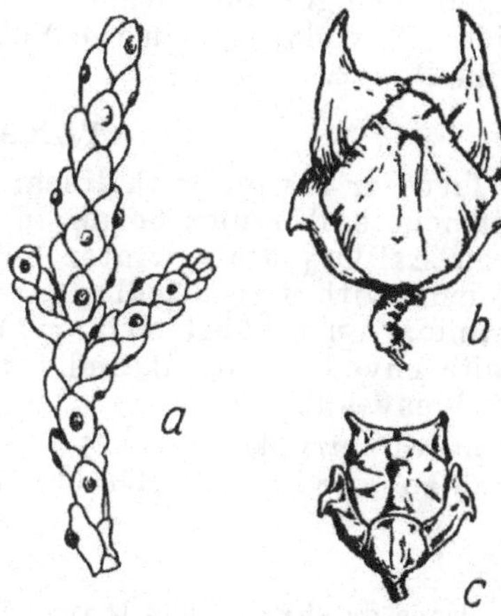

Fig. 102. *a*, Cupressus macnabiana Murr.; branchlet x 6; *b*, cone x 1; *c*, var. bakeri Jepson; cone x 1.

5. JUNÍPERUS L. JUNIPER

Trees or shrubs. Leaves in whorls of 3 or opposite, scale-like, imbricated, closely appressed and adnate to the branchlets, or linear-subulate and spreading. Stamens and ovules on separate trees. Staminate catkins with many stamens, each with 2 to 6 pollen-sacs. Ovulate catkins of 3 to 6 succulent coalescent scales, each bearing 1 or 2 ovules. Cones fleshy and berry-like, ripe in the second year, in ours 1 to 3-seeded; seed-leaves 2 to 6. (Ancient Latin name.)

Catkins axillary; leaves linear or lanceolate, acute, cuspidate, spreading, white-glaucous above, 3 to 6 lines long; subalpine shrub....................1. *J. communis*.

Catkins terminal on short branchlets; leaves ovate, scale-like, ½ to 1 line long, closely appressed to the branchlets in whorls of 3, with a more or less distinct dorsal gland or pit.

Berries reddish brown, oblong; cotyledons 4 to 6; medium altitudes..2. *J. californica*.

Berries blue-black, globose or subglobose; cotyledons 2; high montane............ 3. *J. occidentalis*.

1. **J. commùnis** L. var. **montàna** Ait. DWARF JUNIPER. Low or prostrate shrub, ⅓ to 1 ft. high or less, forming patches a few feet in diameter; leaves rigid, 3 (rarely 2) at a node with very short internodes; berries globose, blue, covered with white bloom, 1½ to 2½ lines long.—Sierra Nevada, 8000 to 9500 ft. from Mono Pass to Mt. Shasta, thence w. to Del Norte Co.; n. to arctic regions; rare in Cal.

2. **J. califórnica** Carr. CALIFORNIA JUNIPER. Usually a shrub, much-branched from the base, 2 to 15 ft. high, or occasionally a tree up to 40 ft. high; berries reddish or brownish, covered with a dense white bloom, subglobose or oblong, 4 to 8 lines long, with dry fibrous sweet flesh and 1 to 3 seeds; seeds ovate, acute, the thick polished bony shell angled or ridged.—Dry hills or mountain sides, w. Mohave Desert, s. to L. Cal., n. in the Sierra Nevada foothills to Kern Co. and in the inner Coast Ranges to San Benito Co.; local near Coulterville and in the mountains w. and n. of Clear Lake. Var. UTAHÉNSIS Engelm. Branchlets very slender; berries subglobose; often bluish.—Mohave Desert; e. to Utah. (J. utahensis Lemm.)

3. **J. occidentàlis** Hook. SIERRA JUNIPER. Tree 10 to 25 or even 65 ft. high; the trunk 1 to 5 ft. in diameter; bark dull red, shreddy; berries globose to ovoid, blue-black with a whitish bloom, 3 to 5 lines long, almost smooth or minutely umbonate, with resinous juicy flesh and 2 seeds (rarely 1 or 3); seeds flat on the face, the convex back with 3 to 5 resinous-glandular pits.—

Granite ridges and cirques, subalpine at 7000 to 10,500 ft., common in the Sierra Nevada, rare in the Yollo Bolly Mts., San Bernardino Mts. and Panamint Range.

TAXÀCEAE. Yew Family

Trees or shrubs with linear leaves 2-ranked by a twist in their petioles. Stamens and ovules borne on different trees and appearing in early spring from axillary scaly winter buds. Stamens united by their filaments into a column with 4 to 8 pollen-sacs pendent from each filament. Ovule solitary, terminal on a short axillary branch. Seeds set loosely in a fleshy cup, or quite enveloped by it and thus appearing drupe-like, ripe in first autumn; seed-leaves 2.

Fruit red, berry-like, leaves ½ to ⅔ in long, acute . . . . 1 Taxus
Fruit green or purplish, plum-like; leaves 1¼ to 2½ in long, stiffish, bristle-pointed . . . 2. Torreya

1. TÁXUS L. Yew

Trees or shrubs, the leaves bluntish or merely acute. Stamens 7 to 12 in a cluster, the 4 to 9 pollen-sacs borne under a shield-like crest. Ovule seated upon a circular disk which in fruit becomes cup-shaped, fleshy, and red, surrounding the bony seed, the whole berry-like. (Ancient Latin name of the yew, probably from Greek toxon, a bow, the wood used for bows.)

1. **T. brevifòlia** Nutt. Western Yew. Fig. 103. Small tree 10 to 30 ft. high, the crown irregular with the branches of unequal length and standing at various angles but tending to droop; bark thin, red-brown, shreddy; leaves linear, acute at apex, shortly petioled, flat, with midrib in relief above and below, 3 or mostly 6 to 8 lines long, 1 line wide, spreading right and left in flat sprays; seeds borne on the under side of the sprays and when mature set in a fleshy scarlet cup, the whole looking like a brilliantly colored berry.—Deep, shady cañons: Santa Cruz Mts. to Mendocino Co. and northw. to Mt. Shasta, thence s. through the Sierra Nevada to Calaveras Co. Localities few and scattered, and the individuals not numerous. Wood fine and close-grained, hard, heavy and durable.

Fig 103 Taxus brevifolia Nutt.; *a*, fr. branchlet x ½; *b*, long. sect. fruit and seed x ¾.

2. TÓRREYA Arn. Stinking Yew

Trees with rigid sharp-pointed leaves in 2 ranks. Stamen clusters solitary in the adjacent leaf axils, borne on 1-year-old branches, made up of 6 to 8 whorls of stamens, 4 stamens in a whorl, each filament with 4 pollen-sacs without crests. Ovule completely covered by a fleshy aril-like coat, the whole becoming drupe-like in fruit. Seed with thick, woody outer coat, its inner layer irregularly folded into the white endosperm; embryo minute. (John Torrey, Professor

Fig 104 Torreya californica Torr ; fr. branchlet x 1/3.

of Botany in Columbia College, long-time a student of western botany, who traveled in California before the days of the Overland Railroad.)

1. **T. califórnica** Torr. California Nutmeg. Fig. 104. Handsome tree 15 to 50 ft. high, the trunk 1/2 to 3 ft. in diameter; leaves rigid, 1 1/4 to 2 1/2 in. long, 1 1/2 lines wide, flat, dark green above, yellowish green beneath and with two longitudinal glaucous grooves, linear or tapering above middle, bristle-tipped, twisted on their short petioles so as to form a 2-ranked flat spray; fruit elliptical, green or purplish, 1 1/8 to 1 3/4 in. long; flesh thin and resinous; endosperm copious, with irregular incisions filled by the inner coat, giving it a marbled appearance so that in cross-section the seed resembles the true nutmeg of commerce.—Cool shady cañons, 100 to 6000 ft.: Coast Ranges from Mendocino Co. to Santa Cruz Mts.; Sierra Nevada from Tehama Co. to Tulare Co.

DIVISION II

PALM TREES. (Monocotyledons)

Palm or palm-like trees, the trunk simple or sparingly branched. Leaves parallel-veined, borne in a tuft at summit of stem or end of branches. Stem increasing in diameter by irregular growth, not by definite concentric layers. Flowers with the parts in 3s or 6s. Seed-leaf 1.

LILIÀCEAE. Lily Family

Perennial herbs, the stems from bulbs, corms or rootstocks, scape-like with basal leaves, or leafy and branching, with us rarely shrubs or trees. Flowers regular, perfect, the perianth with 6 lobes or 6 distinct segments, the 3 outer nearly like the 3 inner, or very unlike, all often colored alike. Stamens 6, sometimes 3, rarely 4. Ovary superior, 3-celled; style 1. Fruit a capsule, rarely a berry.

1. YÙCCA L. Spanish Bayonet

Trees or shrubs with simple or branched stems. Leaves alternate, linear-lanceolate. Flowers large, in terminal panicles, the perianth segments distinct, nearly equal, withering-persistent. Stamens 6. Fruit a capsule, either dry and dehiscent, or somewhat fleshy and indehiscent. Seeds numerous, in 2 rows in each cell, flat, horizontal, with thin black coat. (Indian name for the Manihot, erroneously transferred to these trees.)

The flowers are incapable of self-pollination, each Yucca species being dependent upon a particular moth or species of Pronuba. The female Pronuba works by night, collecting the pollen from the anthers and rolling it into a little ball; she then flies to the flower of another plant, deposits her egg in the ovary and then, in a manner which seems to indicate that her actions are full of purpose and deliberation, climbs to the style and thrusts the pollen ball far down the stigmatic tube. The larva destroys about a dozen seeds, but even if several larvae develop, many perfect seeds are left.

Plants commonly with distinct trunk; filaments papillate; style short or none; stigmas 3, or 1 and 3-lobed; endosperm ruminated
Trunk tall, at summit branching freely; leaf-margin denticulate . 1 *Y. brevifolia*
Trunk commonly short and simple or shortly branched, leaf-margin not serrate, fibrous-shredding.
Style ½ to 1½ lines long . . . . 2 *Y mohavensis*
Style 6 to 10 lines long . . . . 3. *Y. baccata*
Plants without evident trunk, the rosette of leaves on the ground; filaments glabrous; style slender with capitate stigma; endosperm not ruminated. . .4 *Y. whipplei*

1. **Y. brevifòlia** Engelm. Joshua Tree. Tree 20 to 30 ft. high with an open crown of arm-like branches borne on a columnar trunk; leaves bayonet-like, bluish green, 6 to 9 in. long, the edge with minute teeth, not fibrous; panicle 8 to 14 in. long; stigma sessile, 6-lobed; fruit oblong-ovate, slightly 3-angled, 2 to 4 in. long and 1½ to 2 in. broad.—Widely distributed in the Mohave Desert, often forming extensive groves; n. to Kern River Valley and the Coso Mts.; e. to Utah.

2. **Y. mohavénsis** Sarg. Mohave Yucca. Cactus-like shrub or low tree 6 to 15 ft. high, the trunk simple or shortly branched; leaves bayonet-like, 16 to 24 in. long, widest at middle, tapering to apex and to the abruptly widened base, fibrous on the margin; style very short, stigma 3-lobed, each lobe notched at apex; fruit cylindric, fleshy, 2½ to 4 in. long, 1 to 1½ in. thick, usually constricted about the middle.—Mohave Desert, n. to Monterey Co., s. to San Diego Co. and L. Cal., e. to Nev.

3. **Y. baccàta** Torr. Spanish Bayonet. Very similar to Y. mohavensis; leaf rosettes yellow-green, on the ground, rarely rising above it; leaves 1¼ to 2 ft. long; flowering stem 2 to 3¼ ft. high; flowers (2½ or) 3 to 4 in. long; base of filaments forming fleshy papillae; style much elongated; fruit conical.—Providence Mts., e. Mohave Desert; e. to Col. and N. Mex.

4. **Y. whípplei** Torr. Quixote Plant. Flowering stem 8 to 14 ft. high, the leaves in a basal rosette; leaves narrow, 1 to 1¾ ft. long; panicle 3 to 6 ft. long; flowers creamy-white, 1¼ to 1½ in. long, the perianth segments thinnish; filaments much thickened; capsule short-cylindric or subglobose, 1¼ to 2 in. long.—Chaparral belt of cismontane S. Cal., n. in the Coast Ranges to San Benito Co. and in the Sierra Nevada to Kings River. L. Cal. May-June.

PALMÀCEAE. Palm Family

Commonly trees with fibrous roots and columnar unbranched trunks covered with leaf-scars or the bases of leaf-stalks and bearing a tuft of large leaves at summit. Leaves sharply plaited when young, eventually tearing more or less along the lines of the folds. Flowers minute, commonly monoecious, in ours perfect, borne in a large inflorescence enclosed by a spathe. Perianth in two circles, an outer 3-lobed calyx and an inner 3-parted corolla. Stamens 6, inserted on the corolla-tube. Carpels 3, separate or united, each 1-ovuled. Fruit a berry, drupe or nut.

1. WASHINGTÒNIA Wendl. Fan Palm

Trees with fan-shaped much folded blades and long petioles armed with stout hooked spines along the margins. Pistil 1; ovary 3-celled; style and stigma 1. Fruit a berry. (In honor of President Washington.)

1. **W. filifera** Wendl. California Fan Palm. Columnar tree 20 to 75 ft. high, the trunk 1 to 3 ft. in diameter at the enlarged base, covered with a scaly rind and sometimes clothed quite to the ground with a thatch of dead persistent recurved leaf-bases; leaves fan-shaped, 3 to 6 ft. long, with 40 to 60 folds, torn nearly to the middle, the divisions copiously fibrous; petioles 2 to 5 ft. long, very stout; flowers borne in a branched panicle on long

stems, the whole 8 to 12 ft. long; berries borne on pedicels 1 to 1½ lines long, black, oval, 3 to 3½ lines long, with thin flesh surrounding a large seed which is flattened somewhat on the ventral side; endosperm horny.—Westerly and northerly sides of the Colorado Desert, on or above the old beach line of the one-time interior sea, always in moist spots or oases, from near sea level to 3500 ft.

DIVISION III

BROAD-LEAVED TREES. (Dicotyledons)

Deciduous or evergreen trees, the trunk freely parting into branches, rarely persistent through crown as a continuous axis. Leaves netted-veined. Stem increasing in diameter by annual concentric layers of wood laid down inside the bark. Flowers with the parts in 4s or 5s, the perianth commonly differentiated into calyx and corolla, sometimes absent. Seed-leaves 2. Vegetative reproduction by crown-sprouting very common.

BETULÀCEAE. Birch Family

Deciduous trees or shrubs with alternate simple petioled leaves and caducous stipules. Flowers monoecious, in catkins, flowering in late winter before the leaves appear. Staminate catkins elongated, pendulous, falling after flowering; flowers 3 in the axil of each scale; calyx present. Pistillate catkins relatively short; calyx none; ovary 2-celled, one ovule in each cell; styles 2. Fruit a 1-celled 1-seeded nutlet.

Scales of the catkin consisting of a bract and 4 bractlets united, in the woody cone obscurely 5-lobed at apex; pistillate catkins developing into oval or ovoid woody cones which are drooping or spreading and eventually fall whole; stamens 2, 3 or 4 (or 1 to 7)........ ....1 Alnus

Scales of the catkin consisting of a bract and 2 bractlets united, in the fruiting catkin plainly 3-lobed at apex; pistillate catkins solitary, cylindrical and erect in fruit, falling to pieces when mature; stamens 2......................2. Betula.

1. ÁLNUS Hill. Alder

Catkins 2 to 5 in a cluster, usually borne on a forked peduncle, sometimes sessile. Scales consisting of a bract and 4 bractlets united. Staminate catkins pendulous; calyx 4 (or 6)-parted; stamens 1 to 4. Pistillate catkins erect, spike-like, ripening into woody cones, the scales persistent on the axis; flowers 2 in the axil of each scale. Cones spreading or pendulous when mature. Nutlet with a narrow acute margin. (The Latin name.)

Catkins appearing in the early autumn as rather conspicuous naked buds and flowering in the late winter or early spring before the leaves appear; peduncles of the pistillate catkins naked, their branches ½ in. long or less; stamens 1 to 4.
- Trees 30 to 80 ft high; mostly of low altitudes.
  - Leaf-margin plane, with small scattered glandular teeth; bracts of staminate catkin obtuse; stamens 2, sometimes 3, 1, or 4......1 *A. rhombifolia*
  - Leaf-margin coarsely toothed, the entire margin with a narrow underturned edge, bracts of staminate catkin acute or acutish; stamens 4, rarely 3... .. 2. *A. rubra*
- Shrubs 8 to 15 ft. high; leaf-margin coarsely toothed and again finely toothed; stamens 4 or 2; high montane........................3. *A. tenuifolia.*

Catkins appearing in the spring from scaly buds at the same time as the leaves; peduncles of the pistillate catkins leafy (at least at base), their branches ½ to 1 in. long; stamens 4; leaf-margin sharply or laciniately toothed; high montane shrub..... 4. *A. viridis.*

1. **A. rhombifòlia** Nutt. White Alder. Tree 30 to 115 ft. high; bark whitish or gray-brown; leaves oblong-ovate or -rhombic, tapering more or less to base and apex, 2 to 4 in. long; cones ovoid, 5 to 9 lines long.—Banks of rivers and living streams: Coastal S. Cal.; Sierra Nevada cañons; Great

Valley; Coast Ranges except in narrow coast strip occupied by Red Alder; n. to Wash.

2. **A. rùbra** Bong. RED ALDER. Fig. 105. Tree 30 to 90 ft. high; bark very white or white mottled; leaves 2 to 6 in. long, elliptic-ovate, often rusty beneath, the coarse teeth again finely toothed; cones oblong-ovoid, ¾ to 1⅛ in. long—Deep cool cañons or moist flats along the coast: Santa Inez Mts. to Del Norte Co.; n. to Alas. Abundant from Marin Co. to Humboldt Co., where it forms pure groves of singular beauty in marshy bottoms near the sea. Also called Oregon Alder.

3. **A. tenuifòlia** Nutt. MOUNTAIN ALDER. Shrub or small tree, 8 to 14 ft. high; leaves roundish to ovate, thickish at base, truncately rounded (or even subcordate) to cuneate, coarsely toothed and again finely serrate, 1 to 3 in. long; staminate catkins 3 or 4 in a cluster, 3 in. long; pistillate catkins 3 to 8 in a cluster, sessile or with peduncles almost 2 lines long; cones small, 3 to 7 lines long.—Thickets on wet hillsides or in moist hollows, 5000 to 7000 ft.: Sierra Nevada from Lake Tahoe to Mt. Shasta, thence w. to Trinity Summit and the Siskiyous.

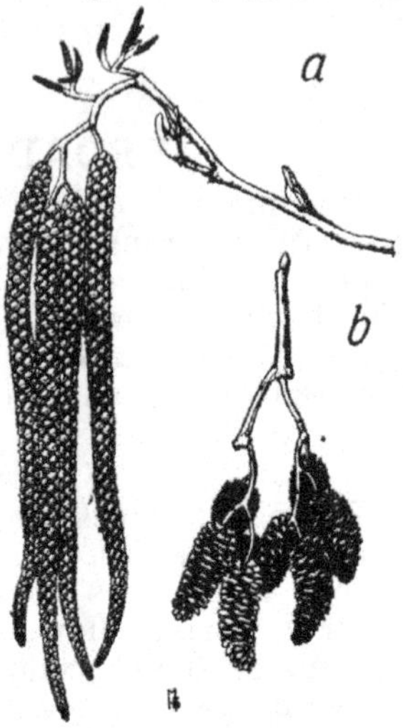

Fig 105 Alnus rubra Bong , *a*, fl branchlet; *b*, cluster of cones. x ¼.

4. **A. víridis** DC. var. **sinuàta** Regel. THIN-LEAF ALDER. Slender shrub 6 to 10 ft. high; leaves broadly ovate, thin, gummy when young, bright green, sharply or laciniately toothed, 2¼ to 3 in. long; catkins appearing in spring at the same time as the leaves, the peduncles of the pistillate leafy at base, their branches ½ to 1 in. long; stamens exceeding the sepals, the filaments long.—High montane, 6500 to 7000 ft.: Trinity Summit to Mt. Shasta; n. to Alas.; Siberia.

2. BÉTULA L. BIRCH

Scales of the catkins consisting of a bract and 2 bractlets united. Flowers 3 in each axil or sometimes 2 in the pistillate catkin. Staminate catkins 1 to 3 in a cluster, sessile or short-peduncled, pendulous; calyx 4 (or 2)-lobed; stamens 2, each filament with 2 distinct anther-cells. Pistillate catkins solitary on the peduncle and erect. Cones erect, long and slender, the scales falling away from the axis (on the tree) when the fruit is mature. Nutlet seed-like, with a broad thin wing. (Latin name of the birch.)

Leaves 1 to 2 in long; lobes of bracts broad, usually parallel, acutish . .1 *B occidentalis*
Leaves ½ to 1 in. long, lobes of bracts narrow, divergent, obtusish . . . .2 *B. glandulosa*

1. **B. occidentàlis** Hook. WATER BIRCH. Slender tree 10 to 25 ft. high with red-brown smooth bark; leaves round-ovate, sharply serrate, mostly acute at apex, almost or quite glabrous, 1 to 2 in. long; petioles 4 to 5 lines long; staminate catkins 2 to 2½ in. long; pistillate catkins 1½ in. long in fruit, 3 or 4 lines in diameter.—Cañon streams, 2500 to 8000 ft., Sierra Nevada from Bubbs Creek and Inyo Co. to Mt. Shasta, thence w. to Humboldt Co.; n. to B. C.

2. **B. glandulòsa** Michx. SCRUB BIRCH. Shrub 1 to 4 ft. high with glandular-warty twigs; leaves roundish, serrate except at base, ½ to 1 in. long; staminate catkins commonly solitary; pistillate catkins 4 to 9 lines long.—High montane, Lassen and Modoc Cos.; n. to arctic circle.

FAGÀCEAE. Oak Family

Trees or shrubs with alternate simple leaves and promptly deciduous stipules. Flowers monoecious, apetalous, appearing with the leaves in the deciduous kinds. Staminate flowers in catkins; calyx parted into several lobes; stamens 4 to 12. Pistillate flowers 1 to 3 in an involucre of imbricated scales, the involucres borne in reduced or short catkins, often solitary, or sometimes 1 or 2 at base of staminate catkin; ovary adherent to the calyx, 3-celled, 6-ovuled, only one ovule maturing, the remaining ovules and the other two cells abortive. Fruit a nut borne singly in a scaly cup or 1 to 3 in a spiny bur.

Fruit an acorn; catkins simple
- Catkins unisexual, the staminate drooping . . . .1 Quercus
- Catkins erect, all with staminate flowers, pistillate flowers at base of some of them . 2 Lithocarpus

Fruit a spiny bur; catkins erect, often branching, unisexual, or with pistillate flowers at base of some of the staminate catkins ....3. Castanopsis

1. QUÉRCUS L. Oak

Trees or shrubs of slow growth, hard wood and usually contorted branches. Flowers greenish or yellowish. Staminate catkins pendulous, one or several from the lowest axils of the season's shoot. Pistillate flowers borne in the upper axils of the season's shoot, the ovary with 3 to 5 styles or stigmas. Fruit an acorn, the nut set in a scaly cup. Abortive ovules often discernible in the ripe or nearly ripe acorn. (Latin name of the oak.)

A. White Oaks.—Bark commonly white or whitish, wood light-colored; stamens mostly 6 to 9; stigmas sessile or nearly so; abortive ovules mostly towards base of nut.

1 Acorns maturing the first autumn; nut glabrous on the inner surface.

Deciduous species.
- Branchlets pendulous; acorn cups deep, the nut long and slender; leaves pinnately parted with coarsely 2 to 3-toothed lobes; trunk bark dark brown, deeply cuboid checked; valleys ... . 1 *Q lobata*
- Branchlets not pendulous; acorn cups shallow; trunk bark white, shallowly checked but smoothish
  - Leaves dark lustrous green above, rusty or pale beneath, 5 to 7-parted; nut subglobose or oblong-cylindric; mossy trees; mainly North Coast Ranges . 2. *Q garryana*
  - Leaves bluish green above, pale beneath, oblong, coarsely toothed or entire; nut oval, often swollen at or below middle; interior dry foothills ...... 3. *Q. douglasii*

Evergreen species.
- Small tree; leaves blue-green, oblong, mainly entire; nut subcylindric; S Cal .... . 4. *Q. engelmannii*
- Shrubs; cups saucer-shaped
  - Branches rigid; leaves ¾ to 1 in long; chaparral areas
    - Leaves brittle, plane, light green, oblong, spiny-dentate or entire; nuts oval to cylindric, blunt or pointed . ... . .5 *Q dumosa*
    - Leaves tougher, dark green, convex above, regularly dentate; nuts short cylindric or subglobose, very obtuse . . .6 *Q durata.*
  - Branches slender, pliable; leaves 3 to 4½ in long, strongly parallel-nerved, toothed, chestnut-like; local in Siskiyous and vicinity..7 *Q sadleriana*

2 Acorns maturing the second autumn; nut tomentose or hairy within.

Trees; acorn cup usually very large and thick
- Leaves 2 to 3½ in long, densely woolly when young, with prominent regular parallel nerves; islands off south coast ....8 *Q. tomentella*
- Leaves commonly 1 to 2 in long, entire or spinose-toothed, dull green above, lead-color beneath or with a golden fuzz when young; cup typically like a yellow turban; mountains...... ....9. *Q chrysolepis*

Shrubs; acorn cup sub-turbinate or low bowl-shaped, thin; leaves ½ to 1½ in long
- Branches slender and pliable, forming broom-like tufts at top of stems; leaves mostly entire, no golden fuzz, high montane . . .. .10. *Q vaccinifolia.*
- Branches rigid, spreading; leaves dentate-prickly, olivaceous above, pale beneath; S and L Cal ....11 *Q palmeri*

B. Black Oaks.—Bark dark or black, wood dark or reddish; stamens 4 to 6; stigmas on long styles; abortive ovules mostly towards top of nut; nut tomentose within.

Acorns maturing the first autumn; nut slender-ovate; leaves roundish or elliptic, convex above; coast valleys and hills.............. .. . .12. *Q. agrifolia*
Acorns maturing the second autumn.
Leaves oblong, obtuse or tapering to the acute apex, plane, pale yellowish below; nut slender ovate, often streaked longitudinally; interior valleys and hills..... 13. *Q. wislizenii*
Leaves pinnately parted or toothed, always bristle-tipped; nut oblong, obtuse; mountains
Deciduous tree; leaf divisions generally coarsely toothed; common.14. *Q kelloggii*
Semi-evergreen tree, leaf divisions smaller, generally entire; rare.15 *Q morehus*

1. **Q. lobàta** Neé. VALLEY OAK. Fig. 106. Graceful tree, commonly 40 to 125 ft. tall; trunk 2 to 10 ft. in diameter; bark thick, cuboid-checked; leaves 3 to 4 (rarely 6) in. long, 2 to 3 in. broad, green above, paler beneath, yellow-veined, parted to the middle or nearly to the midrib into 3 to 5 pairs of lobes; lobes most commonly broadened towards the end, less frequently pointed, coarsely 2 or 3-toothed at apex or sometimes entire; cup drab-brown, deeply hemispherical, very warty, ½ to ¾ in. deep; nut long conical, at first bright green, later mahogany or chestnut-brown, 1½ to 2¼ in. long, ½ to ¾ in. in diameter.—Rich soils of the Sacramento and San Joaquin valleys, Sierra foothills and Coast Range valleys, but not in valleys facing the sea; s. to San Fernando Valley. The most characteristic tree of well-watered valley floors, the round-topped crown typically broader than high, its spreading branches finally ending in long cord-like branchlets which sometimes sweep the ground (whence "Weeping Oak"). Called Water Oak, White Oak and Mush Oak by settlers and Roble by the Spanish-Californians. Var. WÁLTERI Jepson. Leaves roundish in outline, sinuses very narrow; nuts very thick.—Kaweah River basin at about 4000 ft. Var. TURBINÀTA Jepson. Leaves large and thicker than in the species; nuts turbinate.—Little Lake Valley, Mendocino Co.

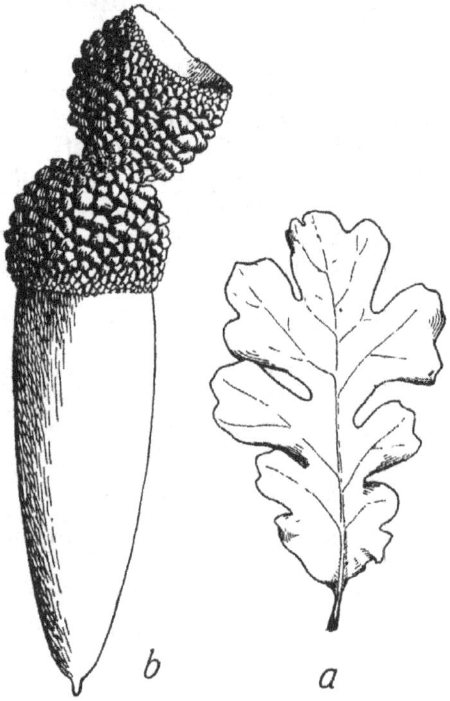

Fig 106 Quercus lobata Neé; *a*, leaf x ½; *b*, acorn x 1.

2. **Q. garryàna** Dougl. OREGON OAK. Fig. 107. Round-headed tree 25 to 55 ft. high, the trunk bark white, thin, superficially checked into small squarish scales; leaves 3 to 4 (or 6) in. long, 1½ to 4½ in. wide, dark lustrous green and nearly glabrous above, rusty or pale, finely pubescent and yellow-veined beneath, leathery in texture and pinnately parted into 5 to 7 (rarely 9) lobes with mostly deep and often acute sinuses; lobes entire or with 2 or 3 coarse rounded unequal teeth; cup very shallow, 6 to 9 lines broad, with tuberculate scales; nut bulging beyond the small cup, typically subglobose but varying

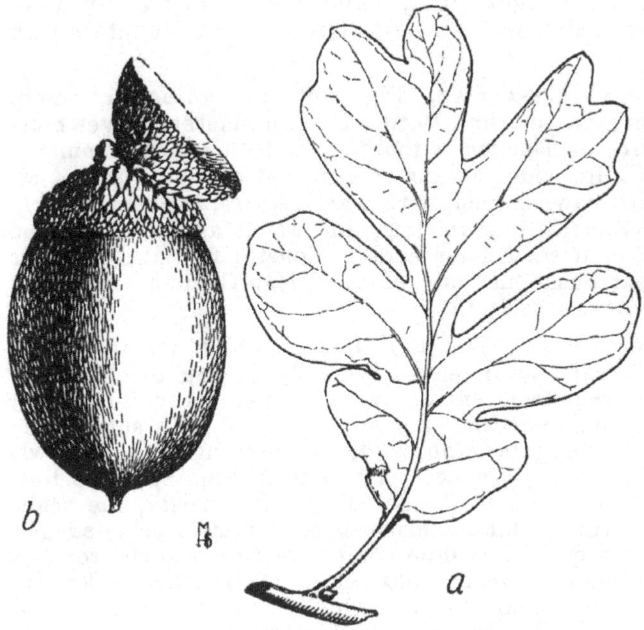

Fig. 107. Quercus garryana Dougl ; *a*, leaf x ½ ; *b*, acorn x 1.

to obovoid or subcylindric, although always rounded at apex, ¾ to 1 in. long, ⅔ to ¾ in. thick, its surface polished and shining.—Rich mountain slopes: Santa Cruz Mts. and Mt. Tamalpais to Mendocino and Humboldt Cos.; far n. to B. C. Var. BRÈWERI Jepson. Brewer Oak. Scrub form; leaves 1 to 2 in. long. — High altitudes from Yollo Bolly Mts. to the Siskiyou Mts. Var. SEMÒTA Jepson. Kaweah Oak. Shrub; leaves 3 to 4 in. long.—Chaparral belt of the s. Sierra Nevada.

3. **Q. douglásii** H. & A. BLUE OAK. Fig. 108. Tree 20 to 60 ft. high, the white trunk bark shallowly checked into small thin scales, this characteristic roughness extending well out to the smaller branches; leaves minutely pubescent, bluish green above, pale beneath, 1 to 3 in. long, ½ to 3 in. wide, mostly

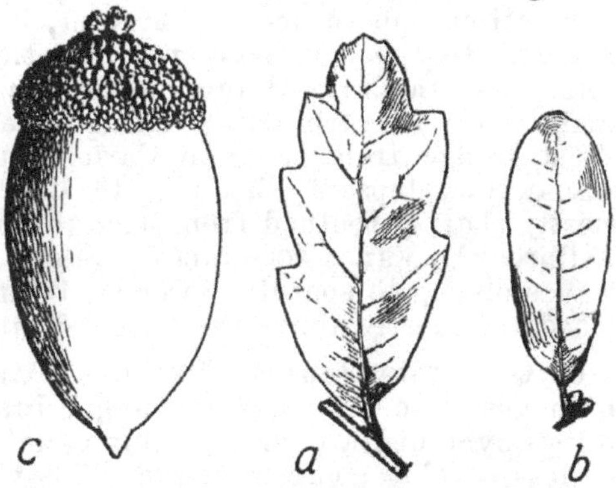

Fig 108 Quercus douglasii H. & A.; *a*, *b*, leaves x ½ ; *c*, acorn x 1.

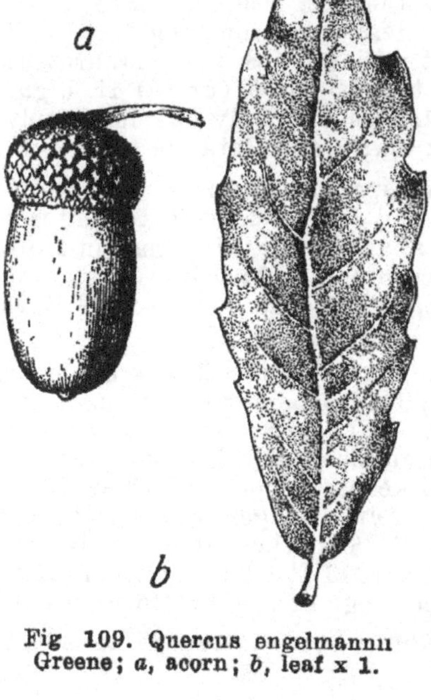

Fig 109. Quercus engelmannii Greene; *a*, acorn; *b*, leaf x 1.

oblong to obovate, entire, or coarsely and often unequally few-toothed, or shallowly lobed; cup 4 to 6 lines broad, of less diameter than the nut and very shallow, the scales developing small wart-like processes; nut ¾ to 1½ in. long, 6 to 10 lines thick, dark or light brown, oval in outline but variable, often much swollen just below or at the middle or only on one side, or again narrow and tapering to apex.—Dry foothills, common, forming open stands: Sierra

Nevada foothills; inner Coast Ranges; approaching the ocean only in Monterey Co., where the Redwood Belt is narrowest. Often called Mountain Oak and Iron Oak by settlers.

4. **Q. engelmánnii** Greene. MESA OAK. Fig. 109. Spreading or round-crowned tree 15 to 50 ft. high with trunk ½ to 3 ft. in diameter; leaves blue-green, oblong, obtuse, entire or sometimes toothed, ¾ to 3 (most commonly 1¼ to 1¾) in. long, ⅝ to 1 in. wide; acorns ripe in first autumn; cup ¾ in. broad, shallow or sometimes bowl-shaped, with warty scales, inclosing nearly ½ the nut which is subcylindrical, ½ in. long and about as thick, or 1 in. long, relatively less thick and sometimes acute.—Coastal S. Cal.: foothills of the San Gabriel Mts. near Pasadena; mts. of San Diego Co. in a zone about 15 to 40 miles back of the sea; s. to L. Cal.

5. **Q. dumòsa** Nutt. SCRUB OAK. Fig. 110. Shrub 2 to 8 ft. high, with tough rigid branches and branchlets; leaves typically oblong to elliptic or roundish, entire or more commonly irregularly spinose-serrate, or sinuate-lobed with sharply cut or angular sinuses, ¾ to 1 in. long; cup shallowly or deeply saucer-shaped to turbinate, 5 to 8 lines broad, 2 to 5 lines deep, often rusty, the scales tuberculate, sometimes so regularly as to suggest a quilted cushion; nut oval to cylindric, rounded or pointed at apex, ¾ to 1⅛ in. long.—Common chaparral shrub in the mts. of S. Cal., ranging northw. through both the Coast Ranges and Sierra Nevada, more or less abundant in the middle and southerly parts of those ranges, rarer in the north. It is highly variable in leaf texture and outline and in acorn character, both of cup and nut. Var. TURBINÉLLA Jepson. Grey Oak. Leaves pale, brittle; cups small, turbinate; nuts slender, pointed.—San Carlos Range, s. to L. Cal. Var. ALVORDIÀNA Jepson. Leaves similar to var. turbinella; nuts slender, very long (1⅝ in.).—Inner South Coast Range from the San Carlos Range to the San Emigdio Mts. Var. ELEGÀNTULA Jepson. Shrub 10 to 15 ft. high; leaves narrow, regularly and mostly sharply toothed from base to apex, but variable.—S. Cal. (Monrovia; Fallbrook). Var. MACDONÁLDII Jepson. Small tree 15 to 35 (or 50) ft. high; leaves oblong to spatulate-oblong, 1¼ to 2¾ in. long, sharply but not deeply lobed and mostly above the base.—Santa Cruz Isl.; Santa Catalina Isl.

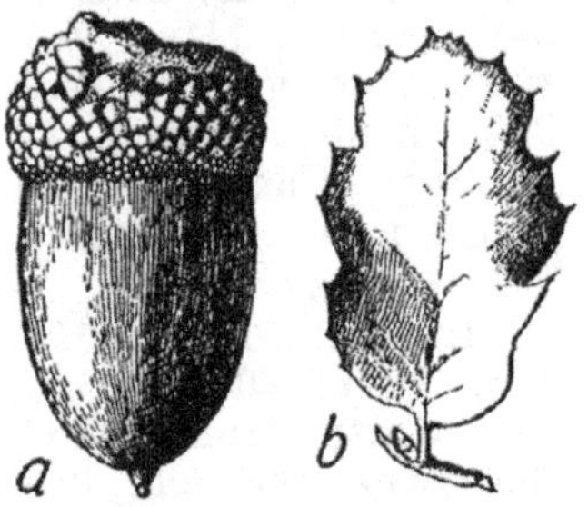

Fig 110 Quercus dumosa Nutt.; *a*, acorn; *b*, leaf x 1.

6. **Q. duràta** Jepson. LEATHER OAK. Low spreading shrub with rigid branches, 2 to 5 ft. high; younger branches and leaves densely tomentose; leaves oval, dentate with prickly equal teeth, above convex, the margin more or less revolute; cup bowl-shaped, 8 to 9 lines broad, 4 to 5 lines high, the scales tuberculate; nut short, thick, cylindric, rounded at apex, 7 to 9 lines long.—Monterey Co. to Napa Co.

7. **Q. sadleriàna** R. Br. Campst. DEER OAK. Shrub mostly 2 or 3 but even 8 ft. high with several stems from the base; leaves persistent through the winter and until after the new leaves appear in the next summer, oblong-ovate, 3 to 4½ in. long, the lateral nerves prominent, regular and parallel; stipules oblanceolate, ½ to ¾ in. long, fur-like on account of their dense covering of rusty hairs and persisting as long or even longer than the leaves; acorns ripe in first autumn; cup inclosing about ⅓ of the nut which is oval and about ¾ in. long.—High montane, 5000 to 7000 ft.: Trinity Summit to the Siskiyou Mts.; sw. Ore. Most restricted in range of any Californian oak.

8. **Q. tomentélla** Engelm. ISLAND OAK. Round-headed tree 25 to 40 ft.

high; leaves elliptic to oblong, tomentose or glabrate and light green above in age, strongly parallel-nerved beneath, 2 to 3½ in. long; cup 1 to 1½ in. wide, ½ to ¾ in. deep, its scales imbedded in a dense tomentum but the tips free; nut subglobose, bluntish, 1 in. long.—Santa Cruz, Santa Rosa, Santa Catalina and San Clemente islands of the Santa Barbara group; also Guadaloupe Isl. It is a strictly insular species.

9. **Q. chrysólepis** Liebm. MAUL OAK. CAÑON OAK. Fig. 111. Tree 20 to 60 ft. high with roundish or often spreading crown; trunk bark whitish, rather smooth; leaves 1 to 2 (sometimes even 4) in. long, thick, green above, yellow beneath with a fine fuzz or powder, or eventually lead-color or dull white, ovate or oblong-ovate, acute at apex, entire, or with entire and toothed leaves frequently found on the same twig; typical cup thick and round-edged with a fine fuzzy or felt-like tomentum concealing the scales, the whole suggesting a yellow turban; nut ovate, globose, or cylindric, rounded at apex or sharply pointed, 1 to 1¼ in. long, ¾ to 1 in. thick.—Mountain ridges, slopes and cañons almost throughout Cal., extremely variable in habit and in the acorns. Not occurring in the foothills. Called also Gold-cup Oak and Mountain Live Oak. It furnishes the most valuable wood amongst our species, being strong, tough and close-grained. Var. NÀNA Jepson. Leaves ovate, acute, small (¾ to 1½ in. long).—Scrub form in the chaparral. Var. HANSÉNII Jepson. Low tree; nut slender-cylindric.—Amador Co. Var.

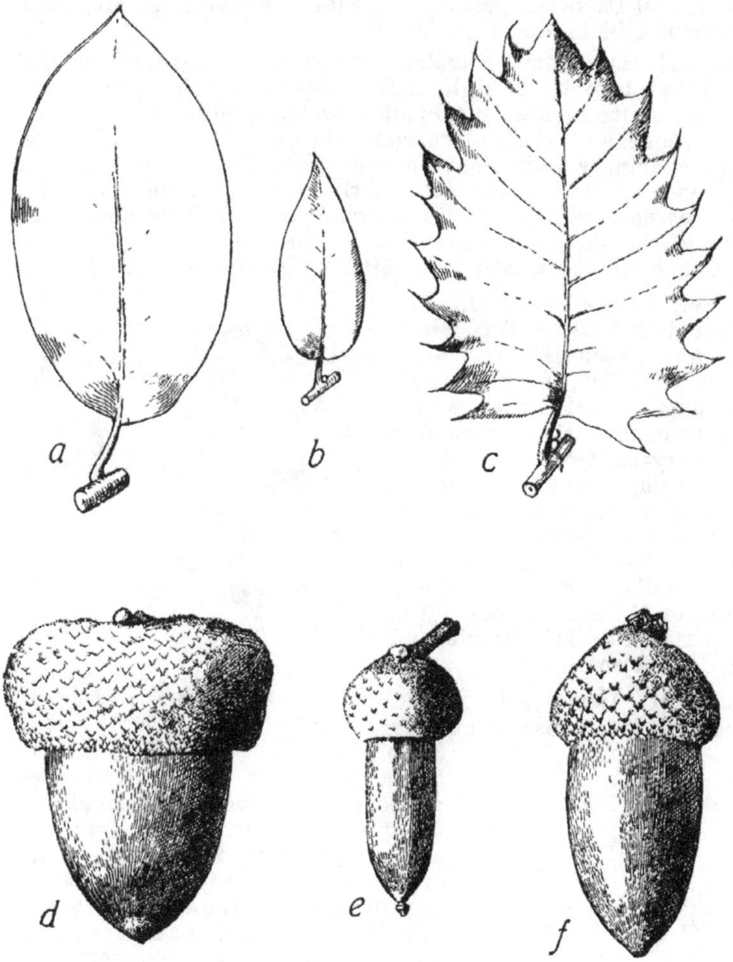

Fig 111 Quercus chrysolepis Liebm.; *a*, *c*, typical leaves; *d*, acorn; *b*, var. nana Jepson, leaf; *e*, var. hansenii Jepson, acorn; *f*, var. grandis Jepson, acorn. x 1.

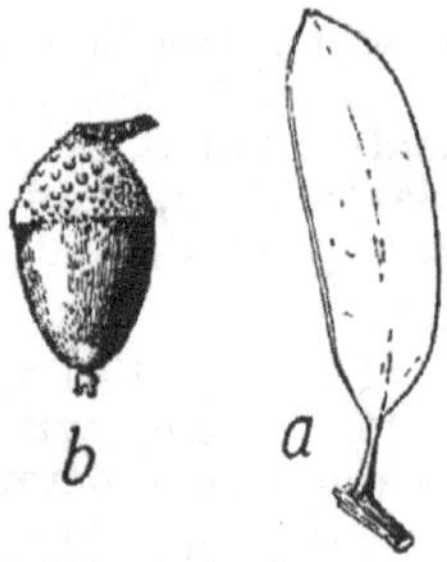

Fig. 112 Quercus vaccinifolia Engelm; *a*, leaf; *b*, acorn. x 1.

GRÁNDIS Jepson. Tree 60 to 110 ft. high, usually with very tall trunks; cup rather thin, the pubescence thin, not concealing the scales.—Deep cañons, Mendocino Co.

10. **Q. vaccinifòlia** Engelm. HUCKLEBERRY OAK. Fig. 112. Low evergreen shrub; branches slender and pliable, forming broom-like tufts at top of stems; leaves very small, mostly entire, no golden fuzz; acorn-cup thin; nut globose-ovate, 4 to 6 lines long.—High montane, 5000 to 10,000 ft., often gregarious: Sierra Nevada; Trinity Mts. to the Siskiyou Mts.

11. **Q. pálmeri** Engelm. PALMER OAK. Rigidly branched shrub 5 to 15 ft. high; leaves typically elliptic to roundish ovate or nearly orbicular, wavy-spinose, ½ to 1½ in. long, undulate, coriaceous and stiff, olivaceous above, pale or whitish beneath; young leaves sparingly pubescent on the upper surface and with a dense but thin yellowish or later white felt on the lower surface; cup thinnish, sub-turbinate but shallow, 5 to 7 lines broad, 3 to 5 lines deep; nut ovate, ¾ to 1 in. long, the shell tomentose within; seed-leaves purple, separable.—Mts. of S. Cal. from the Santa Rosa Mts. to Jacumba; s. to L. Cal.

12. **Q. agrifòlia** Neé. COAST LIVE OAK. Fig. 113. Low broad-headed tree 20 to 70 ft. high; trunk bark smooth and beech-like or irregularly fissured; leaves roundish, elliptic, sometimes ovate or oblong, usually with spine-tipped teeth or sometimes entire, commonly 1 or 2 in. long but varying from ½ to 4 in., usually convex above; cup broadly turbinate, 4 to 7 lines deep, usually embracing only the base of the nut; nut slender, pointed, 1 to 1½ in. long, 5 to 7 lines thick.—Abundant on rich Valley floors, rocky hills and steep cañon sides from cismontane S. Cal. through the Coast Ranges as far north as Mt. Diablo and Napa and Sonoma Cos. Also L. Cal.

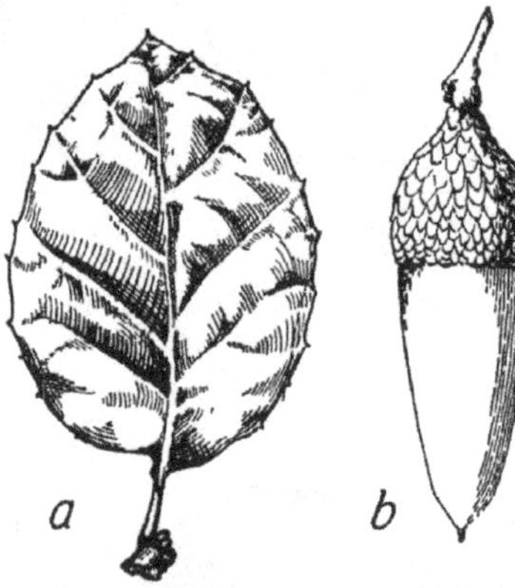

Fig 113 Quercus agrifolia Neé; *a*, leaf; *b*, acorn x 1.

13. **Q. wislizènii** A. DC. INTERIOR LIVE OAK. Fig. 114. Round-headed tree most commonly 30 to 75 ft. high; trunk 1 to 3 ft. in diameter with a thick brittle bark which is very smooth or sometimes roughly fissured; leaves typically oblong (varying to elliptic, ovate or ovate-lanceolate), either tapering to apex or rounded, 1 to 2½ (or 4⅔) in. long, glabrous, green and shining above, pale yellowish green below, the margin entire or spiny-toothed; cup deeply cup-shaped to hemispherical, embracing ¼ to ½ the nut, 6 to 7 lines broad, its scales thin, red-brown; nut cylindric and tapering to the apex or conical, often longitudinally banded with dark lines converging at the summit, 1¼ to 1⅝

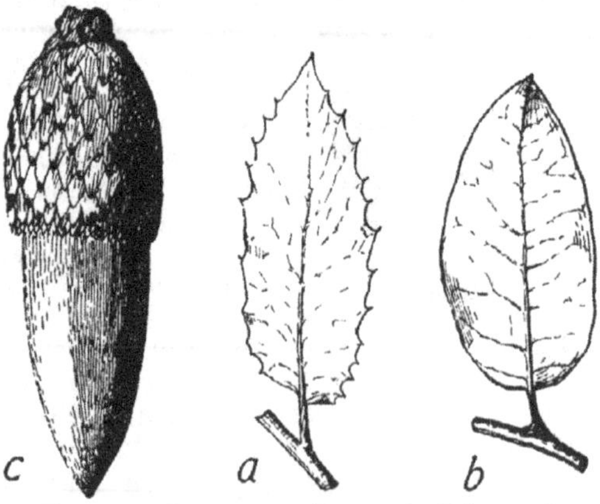

Fig 114. Quercus wislizenii A DC; *a*, *b*, leaves; *c*, acorn. x 1.

in. long.—Hill slopes or moist valley levels, 200 to 5000 ft.: Great Valley; Sierra Nevada foothills from Kern Co. to Shasta Co., thence s. in the inner Coast Ranges to the Vaca Mts. and to Ukiah Valley. Var. FRUTÉSCENS Engelm. Scrub form in the chaparral usually 4 to 6 ft. high; branchlets very rigid; leaves small, thick.—Coast Range summits and ridges w. to the ocean and s. to S. Cal. and L. Cal.

14. **Q. kellóggii** Newb. CALIFORNIA BLACK OAK. Fig. 115. Graceful tree with broad rounded crown, 30 to 80 ft. high; trunk bark dark, checked into small plates; leaves deeply and mostly sinuately parted, with about 3 lobes on each side ending in 1 to 3 or more coarse bristle-tipped teeth, lustrous green above, lighter beneath, often white with a fine tomentum when young, 4 to 10 in. long, 2½ to 6 in. wide; cup large, ½ to 1 in. deep, ¾ to 1⅛ in. broad, its scales thin, with a membranous and sometimes ragged margin; nut oblong, rounded at apex, deeply set in the cup, 1 to 1¼ in. long and ¾ in. thick, covered at first with a fine fuzz. — Mountain slopes and gravelly valleys, 1500 to 6500 ft., or sometimes ascending to 8000 ft. or descending to 200 ft.: high mts. of S. Cal.; Sierra Nevada; Coast Ranges, but not near the sea; n. to southwestern Ore.

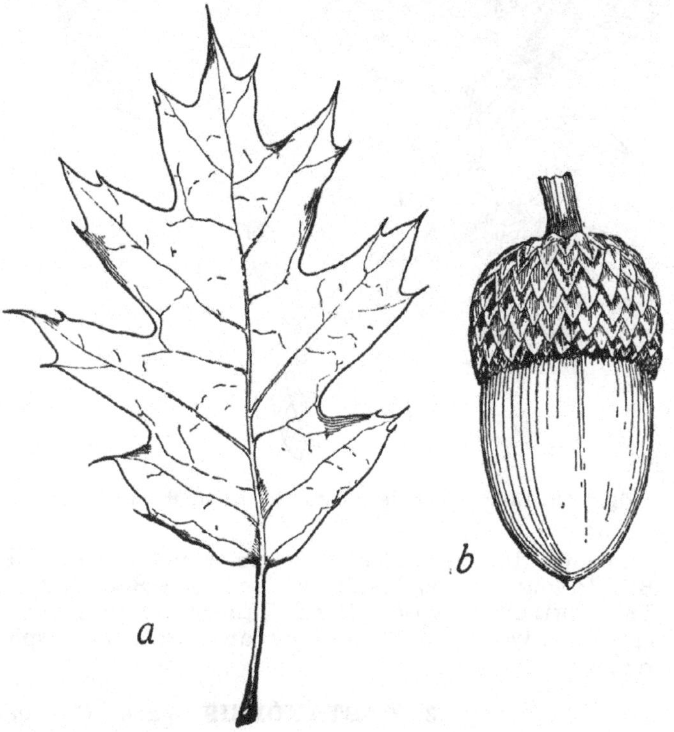

Fig. 115. Quercus kelloggii Newb ; *a*, leaf x ½ ; *b*, acorn x 1.

15. **Q. mórehus** Kell. ORACLE OAK. Tree 25 to 50 ft. high; leaves oblong to elliptic, 2½ to 4 in. long, sinuately but rather shallowly lobed, the lobes pointing upward and spinose-tipped; cups similar to those of Q. wislizenii or more cup-shaped; nuts cylindric, about 1 in. long, 6 to 7 lines thick, minutely pubescent.—Occurring only at scattered stations, the individuals usually few, 200 to about 5000 ft.: Sierra Nevada from Eldorado to Tulare Co.; Coast Ranges from Trinity Co. to Alameda Co. and s. to San Bernardino Mts.

2. **LITHOCÁRPUS** Bl. TAN OAK

Trees or shrubs with evergreen leaves and erect catkins. Staminate flowers one in a place, densely disposed in elongated simple erect catkins; stamens 8 to 10, four times as long as the 5-parted calyx. Pistillate flowers 1 in an involucre, the involucres few at the base of some of the staminate catkins; calyx often with rudimentary stamens; ovary 3-celled. Fruit an acorn, the

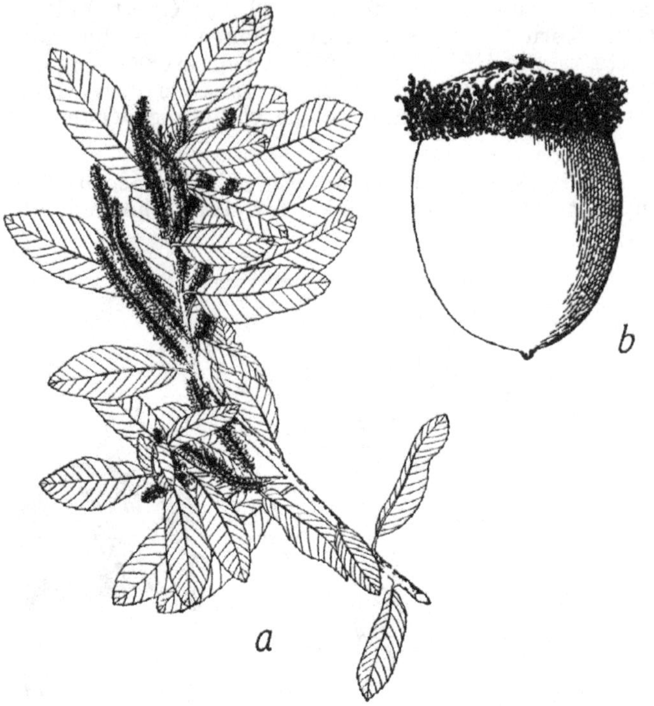

Fig. 116. Lithocarpus densiflora Rehd.; *a*, fl branchlet x ¼; *b*, acorn x 1.

cup with slender spreading scales. (Greek lithos, rock, and karpos, fruit, referring to the acorn.)

1. **L. densiflòra** (H. & A.) Rehd. TAN OAK. Fig. 116. Large tree 50 to 150 ft. high; leaves oblong, acute, strongly parallel - nerved beneath, the nerves ending in the teeth of the margin, 2½ to 4½ in. long, 1 to 1¾ in. wide; catkins slender, 2 to 4 in. long; acorns ripe in second autumn; cup ¾ to 1¼ in. broad; nut globose or short thick cylindric, 1 to 1½ in. long, covered with a deciduous close woolly coat. — Fertile mountain slopes and ridges, Santa Lucia Mts. to Siskiyou Co., thence s. to Mariposa Co. It is associated with the Redwood, or in Mendocino and Humboldt Cos. it becomes most highly developed in the "Bald Hills" country just inside the Redwood Belt. (Pasania densiflora H. & A.) Var. ECHÌNOÌDES Jepson. Scrub Tan Oak. Low shrub 1 to 10 ft. high, its thick entire leaves 1 to 2 in. long and with inconspicuous nerves.—Siskiyou region.

3. CASTANÓPSIS Spach. CHINQUAPIN

Trees or shrubs with evergreen leaves and branchlets lengthening by a terminal bud. Catkins slender, erect. Staminate flowers in clusters of 3, disposed on elongated sometimes branching catkins; calyx 5 or 6-parted; stamens 6 to 12; ovary rudiment present. Pistillate flowers 1 to 3 in an involucre, the involucres on shorter catkins or sometimes scattered at the base of the staminate catkins; calyx 6-cleft with abortive stamens on its lobes; ovary 3-celled with 2 ovules in each cell; styles 3. Fruit maturing in the second season, the spiny involucre inclosing 1 to 3 nuts. Nuts ovoid or globose, more or less angled, usually 1-seeded. (Greek kastanea, chestnut, and opis, resemblance.)

Tree with thick rough bark, less commonly a shrub; leaves 2½ to 5½ in long, usually long-pointed....................... ... .1 *C. chrysophylla*

Shrub, low and spreading, with thin bark, leaves 1½ to 3 in long, usually obtuse....... 2 *C sempervirens*

1. **C. chrysophýlla** A.DC. GIANT CHINQUAPIN. Tree 50 to 115 ft. high with the very thick trunk bark broken into longitudinal furrows; leaves oblong, tapering to base and also to apex (commonly abruptly long-pointed), entire, dark green on the upper surface, at first golden with a fine tomentum

Fig 117. Castanopsis chrysophylla var. minor Benth., fl. branchlet x ½.

below, later light olive-yellow, 2½ to 5½ in. long, ¾ to 1¾ in. wide, the nerves straightish, forking well inside the margins; burs chestnut-like, irregularly 4-valved, containing 1 or sometimes 2 subtriangular nuts 4 or 5 lines long with hard shell and sweet kernel. — Mendocino and Humboldt Cos.; n. to Ore. Var. MÌNOR Benth. Golden Chinquapin. Fig. 117. Shrub 3 to 15 ft. high; leaves trough-like (partly folded along the midrib), very golden below, 2 to 3 in. long. —Rocky ridges and slopes: Monterey to Humboldt Co., mostly near the coast.

2. **C. sémpervirens** Dudley. BUSH CHINQUAPIN. Spreading shrub 1 to 8 ft. high with smooth brown bark; leaves mostly plane, oblong, acutish at base, acute or obtuse at apex, or sometimes tapering upwards from near the base and therefore lanceolate-oblong, 1½ to 3 in. long and 5 to 11 lines wide.—Dry mountain slopes or rocky ridges; high mts. of S. Cal. and Sierra Nevada, 3000 to 8000 ft.; Coast Ranges, 1500 to 4000 ft., but not near the sea.

JUGLANDÀCEAE. WALNUT FAMILY

Deciduous trees with pinnately compound leaves without stipules. Flowers monoecious, appearing after the leaves. Staminate flowers numerous in pendulous lateral catkins. Pistillate flowers few on short erect terminal catkins. Ovary 1 to 3-celled, inferior. Fruit a nut with a dry husk; seed one, deeply 2-lobed.

1. JÙGLANS L. WALNUT

Bark strong-scented. Branchlets hollow, divided into little chambers by pithy partitions. Buds nearly naked. Staminate flower with an irregularly 3 to 6-lobed calyx and numerous stamens. Pistillate flower with a 4-lobed calyx. Seed so lobed as to fit the irregularities of the nut. (Latin Jovis, Jupiter, and glans, nut.)

Leaflets lanceolate; fruits 1½ to 1¾ in in diameter, the nuts smoothish ....1 *J. hindsii.*
Leaflets oblong-ovate to oblong-lanceolate; fruits ¾ to 1 (or 1¼) in in diameter, the nuts with longitudinal channels .. . .2 *J californica*

1. **J. hìndsii** Jepson. CALIFORNIA BLACK WALNUT. Tree 30 to 60 ft. high with erect unbranched trunk 10 to 40 ft. high; crown usually as high as broad; leaflets (2½ or) 3 to 5 in. long.—Central California, on old Indian camp-sites: Walnut Creek; Walnut Grove; Napa Range above Wooden Valley; Gordon Valley.

2. **J. califòrnica** Wats. SOUTHERN CALIFORNIA BLACK WALNUT. Shrub or tree, the stems several from the base or the stem low-branching, forming a wide-spreading crown, much broader than high; leaflets 1¼ to 2½ in. long.—

Coastal S. Cal. from Santa Barbara Co. to the Santa Ana Mts. and e. to the foothill cañons of the San Bernardino Mts. (south slope).

MYRICÀCEAE. SWEET-GALE FAMILY

Shrubs or small trees. Leaves simple, alternate, fragrant, resinous-dotted, without stipules. Flowers in oblong or cylindrical catkins, unisexual, solitary and sessile in the axils of scaly bracts; perianth none. Staminate flower with 4 to 16 stamens; pistillate flower surrounded at base by 2 to 4 small scales or bractlets; ovary superior, 1-celled, 1-ovuled; stigmas 2, filiform, sessile. Fruit a nutlet. Seed without endosperm.

1. MYRÌCA L. WAX MYRTLE

The only genus. (Greek murike, the ancient name of the Tamarisk.)

Evergreen; staminate and pistillate flowers on same shrub; Coast Ranges. 1. *M. californica.*
Deciduous; staminate and pistillate flowers on different shrubs; Sierra Nevada........ 2. *M. hartwegii.*

1. **M. califórnica** Cham. WAX MYRTLE. Thickly branched shrub or small tree, 8 to 30 ft. high; leaves thickish, dark green, glossy, oblong, or oblanceolate-oblong, tapering above to an acute apex, narrowed below to a petiole, 2¼ to 5 in. long, remotely serrate or almost entire; catkins 2 to 5 lines long, the pistillate in the upper, the staminate in the lower, axils; catkins with pistillate flowers above and staminate at base often occur between; stamens 7 to 16, united by their filaments into a cluster longer than the bract; fruit berry-like but really a small globose nut coated with resinous-waxy grains, 2 lines in diameter, the bractlets at the base minute.—Sand-dunes, moist flats or hillsides, Redwood forest slopes, or rocky declivities near the ocean from the Santa Monica Mts. to the Santa Cruz Mts. and Del Norte Co.; n. to Wash. Mar.-Apr.

2. **M. hartwégii** Wats. SIERRA SWEET BAY. Shrub 3 to 5 ft. high; leaves narrowly obovate or oblong, remotely serrulate, puberulent, petioled, 2 (or 1) to 3½ in. long; staminate catkins 1 in. long, the scales ovate-deltoid with scarious margin above the shortly clawed base; stamens 2 to 4, shorter than the scales; pistillate catkins globose, 1 to 1½ lines long, in fruit 5 to 7 lines long, sprinkled with resin globules; bractlets adnate to the sides of the nutlet, enlarged and much thickened in fruit, acutely tipped, 1 line long; nutlet laterally compressed, shorter than the bractlets, not waxy.—Sierra Nevada, 1000 to 4000 ft.: Northfork (Madera Co.); Big Creek near Wawona; Rosasco, Tuolumne Co.; Eldorado Co., and northward to about Yuba Co. May-June. The foliage has an agreeable spicy fragrance.

ULMÀCEAE. ELM FAMILY

Trees or shrubs with alternate simple leaves and fugacious stipules. Flowers perfect or polygamous. Calyx small. Corolla none. Ovary 1-celled; styles or stigma 1. Fruit a samara or drupe.

1. CÉLTIS L. HACKBERRY

Flowers greenish. Calyx 5 or 6-parted, the staminate in cymose clusters at the base of the season's shoot, the pistillate solitary or in pairs in the leaf axils and peduncled. Calyx 5 (or 6)-parted. Staminate flower with 5 (or 6) stamens. Pistillate flower with pistil and 5 (or 6) stamens. Ovary 1-ovuled. Fruit a drupe. (Name of Pliny for a Lotus.)

1. **C. reticulàta** Torr. WESTERN HACKBERRY. Small spreading tree 8 to 12 ft. high; leaves ovate, unequal-sided at the cordate base, serrulate, scaberulous, 1 to 3 in. long on petioles 1 to 3 lines long; flowers 2 lines long; drupe

berry-like, bluish, globose, 3 lines long, on peduncles 2 to 4 lines long.—Near or on western margins of the deserts: Independence; Caliente Creek, Kern Co.; Laguna Mt., San Diego Co.; e. to Col. and Tex., s. to Mex.

SALICÀCEAE. WILLOW FAMILY

Deciduous trees or shrubs of rapid growth, light wood and bitter bark. Leaves simple, alternate, with stipules. Flowers dioecious, arranged in catkins, these falling off as a whole, the staminate after shedding the pollen, the pistillate after ripening of the fruit and dispersion of the seeds. Bracts (or scales) of the catkin scale-like. Calyx and corolla none. Stamens 1 to many. Ovary 1-celled; stigmas 2. Fruit a 2 to 4-valved capsule inclosing many seeds furnished with a tuft of hairs at base.

Scales entire or merely denticulate, persistent or sometimes deciduous; flowers without disk; stamens usually 1 to 5; stigmas short . . . . .1 SALIX
Scales fimbriate or lacerate, caducous; flowers with a broad disk; stamens usually numerous; stigmas elongated or conspicuously dilated . . . . .2. POPULUS

1. SÀLIX L. WILLOW

Trees or shrubs with mostly narrow short-petioled leaves. Winter-buds covered by a single scale. Catkins mostly erect, appearing before or with the leaves; scales entire or merely denticulate, persistent. Staminate flowers with 1 to 9 stamens and 1 or 2 little glands. Pistillate flowers with a gland at the base of the ovary. Stigmas short. (Classical Latin name of the Willow.)

Stamens 3 to 9, their filaments hairy or woolly below; style short; stigmas roundish, subentire; scales pale or yellowish, in the pistillate catkin more or less deciduous by maturity; capsules pediceled; trees, mainly of lower altitudes.
  Petioles with wart-like glands at summit; leaves lanceolate, long-pointed; stipules usually present, roundish; catkins in bud tapering, in flower usually straight, their scales erect . . . . .1 *S lasiandra.*
  Petioles not glandular; stipules usually absent; catkins in bud cylindric.
    Leaves broadly lanceolate, acute, shining green above, usually glaucous beneath; staminate catkins curving, scales reflexed or spreading . .2 *S laevigata.*
    Leaves very narrow, nearly alike on both faces, finely serrulate, often curving towards apex . . . . .3 *S nigra*
Stamens 2 (rarely 1).
  Filaments woolly or hairy below; scales pale, somewhat deciduous; catkins borne on short leafy branchlets, often clustered, leaves linear or lanceolate, shrubs of stream beds at lower altitudes.
    Stigmas linear, raised on a distinct style; ovary densely silky; leaves silvery or green . . . . .4. *S. sessilifolia*
    Stigmas oblong or roundish, sessile; capsule glabrous; leaves green, remotely serrulate . . . . .5 *S. exigua*
  Filaments glabrous; stigmas entire or notched, rarely parted into linear lobes; scales usually black or dark-colored, mostly persistent; trunk bark usually smooth
    Capsules glabrous; leaves dark green above, white-pubescent beneath; catkins sessile, leafless; filaments more or less united; small tree or shrub . . . . 6. *S lasiolepis*
    Capsules tomentose, silky or puberulent.
      Style none; catkins short and dense, their scales black, with white hairs; leaves obovate, usually glabrate . . . . .7 *S. scouleriana*
      Style evident; catkins long and slender; stamens 1, or if 2 the filaments partly united; leaves conspicuously silky beneath. 8 *S sitchensis*

1. **S. lasiándra** Benth. YELLOW WILLOW. Tree 20 to 45 ft. high, the trunk with brown roughly fissured bark; one-winter-old branchlets yellowish; mature leaves lanceolate with long tapering or very slender point, 4 to 7 in. long, ⅝ to 1¼ in. wide; petioles glandular at the upper end; stipules on vigorous shoots conspicuous, orbicular, 5 to 12 lines broad; staminate catkins 1¼ to 3 in. long, usually straight; pistillate catkins 1¼ to 2¼ in. long; scales erect; stamens 4 to 9; ovary glabrous.—Along living streams, valleys and mountains throughout cismontane Cal., 50 to 8500 ft.; n. to B. C.

2. **S. laevigàta** Bebb. RED WILLOW. Tree 20 to 40 ft. high; one-winter-old branchlets reddish; mature leaves oblong-lanceolate to lanceolate, obtusish at base, acute at apex or sometimes long-pointed, serrulate, glabrous, green and shining above, pale or conspicuously glaucous beneath, 2½ to 7½ in. long, ⅝ to 1¼ in. wide; stipules minute and caducous or none; staminate catkins commonly flexuous, 1½ to 4½ in. long; pistillate catkins ¾ to 2 in. long; scales soon spreading or reflexed; stamens 4 to 7 (sometimes 3); ovary glabrous.—Along living streams throughout the state, 100 to 4500 ft. Var. ARAQUÌPA Jepson. One-winter-old branchlets tomentulose.—Dry foothills, Solano Co.

3. **S. nìgra** Marsh. var. **vallícola** Dudley. BLACK WILLOW. Tree 20 to 40 ft. high; leaves lanceolate or linear-lanceolate, long-pointed, often falcate, serrulate, glabrous, green on both surfaces, 2 to 5 in. long, 2 to 3 lines wide; petioles 1 line long; stipules early deciduous; scales obovate, erect; staminate catkins 1½ to 2½ in. long; stamens 3 to 5; pistillate catkins ¾ to 1¼ in. long, in fruit 1 to 2½ in. long, becoming rather lax; ovary scantily pubescent or hoary.—River banks: Sacramento and San Joaquin valleys; Sierra Nevada foothills; S. Cal.; e. to Tex., s. to Mex. (S. gooddingii Ball.)

4. **S. sessilifòlia** Nutt. SANDBAR WILLOW. Shrub with slender stems 5 to 14 ft. high, or becoming a tree up to 25 ft. high; foliage silvery or becoming more or less green; leaves linear, usually tapering to the acute apex and to the narrow but short petiole-like base, entire, 1 to 3 inches long, 2 to 4 lines broad, thinly villous on both surfaces and green, or densely villous and silky, especially on young or sterile shoots; no stipules; catkins on leafy peduncles; staminate catkins ⅓ to 1 in. long, slender (2 lines thick), in bud usually cylindric, the scales with acute green tips; gland long and slender; pistillate catkin ¾ to 1 in. long, 3 lines broad, often not dense; ovary sessile, densely silky; style present, stigmas linear; capsule densely silky, or glabrescent and brown.—Abundant in stream beds: Coast Ranges, Great Valley and Sierra Nevada foothills; n. to Ore.

5. **S. exígua** Nutt. LONGLEAF WILLOW. Shrub 5 to 15 ft. high with bright green foliage; leaves mostly glabrous, or sometimes minutely canescent, lanceolate or linear, tapering to apex and to a short petiole at base, remotely serrulate with cuspidate teeth, ¾ to 5 in. long, 2 to 4 lines wide; catkins terminal on leafy branches; staminate catkins ½ to 1⅛ in. long, 2 lines thick; pistillate catkins ½ to 1 in. long; ovary pediceled or sometimes nearly sessile, glabrous; stigmas very short, sessile; scales densely woolly; fruiting catkin 1¼ to 2½ in. long.—Stream beds, valleys, foothills and mountains throughout the state, 200 to 4000 ft.; far e. (S. longifolia Jepson.)

6. **S. lasiólepis** Benth. ARROYO WILLOW. Shrub or tree 10 to 18 or rarely 35 ft. high, the trunk 3 to 7 in. in diameter with smooth bark or very old trunks shallowly seamed; mature leaves oblong, obovate or linear, acute, obscurely serrulate, dull green and glabrous above, white-pubescent or pale beneath, 1½ to 5 in. long, ⅓ to 1¼ in. broad; catkins appearing before the leaves, sessile, densely silky tomentose in the bud, suberect; scales dark; staminate catkins ¾ to 1½ in. long, 5 to 6 lines thick; stamens 2, filaments glabrous, distinct or united to the middle; pistillate catkins ¾ to 1 in. long, 3 to 4 lines thick, in fruit 1½ to 2¼ in. long; capsule glabrous or puberulent, short-pediceled.—Coast Ranges, Sacramento and San Joaquin valleys and Sierra Nevada foothills, n. to the Klamath River and s. to S. Cal. and L. Cal. The most common willow in the foothill country throughout the state, especially along the summer-dry arroyos but also found on living streams. Also called White Willow. Var. BIGELÒVII Bebb. Leaves broadly obovate or cuneate

oblong, obtuse, entire, ¾ to 1⅜ in. broad; catkins on short leafy peduncles.—San Francisco.

7. **S. scouleriàna** Barr. var. **flavéscens** Schn. NUTTALL WILLOW. Shrub 2 to 15 ft. high or a small tree up to 25 ft. high; leaves broadly obovate or oblong-obovate, entire, rounded at apex or shortly acute, 1 to 1½ (or 4) in. long, ½ to 1¼ in. broad, yellow-green and lustrous above, yellow-veined, glabrate or densely short-silky beneath; catkins appearing before the leaves, oblong or elliptic, ½ to 1 in. long, 5 to 7 lines thick, sessile; scales obovate, rounded at apex, black or black-tipped, covered with white hairs; stamens 2, conspicuously long-exserted, filaments glabrous; ovary white-silky; style none, stigmas broadly linear, sometimes notched at apex; capsule less silky than the ovary.—Montane, 4000 to 10,000 ft.: Sierra Nevada, s. to the San Bernardino Mts., n. to the Siskiyou Mts.; e. to the Rocky Mts. Highly variable. (S. flavescens Nutt.) Var. CRASSIJÙLIS Andr. Branchlets pubescent and tomentose; winter buds pubescent; leaves shorter and broader.—Coast Ranges near the sea. (S. brachystachys Benth.)

8. **S. sitchénsis** Sanson. SITKA WILLOW. Arborescent or shrubby, 5 to 25 ft. high, the trunk 2 to 10 in. in diameter; leaves obovate to oblanceolate, rounded or shortly acute at apex, entire (obscurely serrulate on vigorous shoots), dark green and almost glabrous above, densely tomentose and lustrous silky beneath, 2 to 5 in. long, 1 to 3 in. broad; stipules small, early deciduous or on sterile shoots broad or orbicular, 4 to 6 lines long; staminate catkins 1¼ to 2 in. long, 5 to 6 lines thick; stamens 1, or exceptionally 2 and their filaments more or less united; pistillate catkins ¾ to 2 in. long and 3 lines thick, or in fruit 3 to 5 in. long; scales covered with long white silky hairs, the staminate rounded at apex, the pistillate shorter, broader and more acute; style elongated, stigmas short-oblong, entire or nearly so.—Along the coast from Santa Barbara Co. to Marin and Humboldt Cos.; n. to Alas. Also called Silky Willow.

2. **PÓPULUS** L. POPLAR

Trees with scaly buds and caducous stipules. Leaves rather long-petioled, broad. Winter buds covered by many scales. Catkins appearing before the leaves, in ours pendulous; scales imbricate or lacerate, falling as soon as released by the flowering elongation of the catkin. Stamens inserted on the surface of a concave disk. Ovary seated on a collar-like disk; style short; stigmas 2 to 4, narrow and elongated, or conspicuously dilated. Capsule 2 to 4-valved. Coma of the small seeds long and conspicuous. (Classical Latin name of the Poplar.)

Stamens 40 to 80
- Leaves deltoid-orbicular, broader than long, yellowish green, alike on both faces, valley streams........ .. .. 1 *P fremontii*
- Leaves longer than broad, ovate, dark green above, rusty or silvery beneath; valley and mountain streams .... 2. *P trichocarpa*

Stamens 6 to 12; leaves round-ovate, 1 to 2 in long; high mountains....3 *P. tremuloides*

1. **P. fremóntii** Wats. FREMONT COTTONWOOD. Handsome tree commonly 40 to 90 ft. high with massive crown of spreading branches; leaves triangular or roundish in outline, 2 to 4 in. broad, broader than long, the margin crenate except at the abruptly short-pointed apex and the truncate or subcordate base; staminate catkins 2 to 4 in. long, densely flowered; stamens about 50 to 70; pistillate catkins loosely flowered; stigmas 3 or 4, roundish; capsules on pedicels 2 lines long; seeds copiously provided with long white hairs which soon involve the catkin in a soft cottony mass.—Living streams: Great Valley; Sierra Nevada foothills; South Coast Ranges; S. Cal.; absent or rare in North Coast Ranges.

2. **P. trichocárpa** T. & G. Black Cottonwood. Tree 40 to 125 ft. high with a broad head of upright branches; leaves broadly or narrowly ovate, finely serrate, truncate or heart-shaped at base, acute or tapering to a point at apex, 2½ to 10½ in. long, lustrous green above, rusty-brown beneath when young but at length whitish; staminate catkins 1 to 2 or eventually 5 in. long; stamens 40 to 60 on a slightly one-sided disk; pistillate catkins loosely flowered, 2½ to 3 in. long, in fruit 4 to 10 in. long; stigmas 3, dilated and deeply lobed.—Living streams, mostly 500 to 5000 ft., nearly throughout the state in the greater ranges; n. to Alas.

3. **P. tremuloides** Michx. Aspen. Slender tree with branches gracefully pendulous towards the ends, 10 to 60 ft. high, the trunk 3 to 10 in. in diameter, bark smooth, greenish white, or on old trunks nearly black; leaves round-ovate, finely toothed or almost entire, abruptly tipped at apex with a short sharp point, 1 to 2 in. long; staminate catkins 1½ to 2½ in. long, each flower with 6 to 12 stamens; pistillate catkins 2 to 4 in. long; ovary conical; stigmas 2, very thick below, divided above into 2 slender spreading lobes; style short and thick; seeds minute, brownish, bearing long white hairs.—High montane, 5000 to 10,000 ft.: Sierra Nevada; Mt. San Gorgonio; Trinity Mts.; n. to Alas., e. to Me. and Rocky Mts., s. to Mex.

LAURÀCEAE. Laurel Family

Aromatic evergreen trees and shrubs with alternate simple leaves and no stipules. Flowers perfect, regular. Petals none. Anthers opening by uplifted valves. Ovary superior, 1-celled, 1-ovuled, with a single style. Fruit in ours a drupe.

1. UMBELLULÀRIA Nutt.

Flowers in simple peduncled umbels. Sepals 6. Stamens 9, the three inner with a stipitate orange-colored gland on each side of the filament at base and alternating with scale-like staminodia; anthers 4-celled, 4-valved, the three inner extrorse, the outer introrse. (Latin umbellularia, a little umbel.)

1. **U. califòrnica** Nutt. California Laurel. Tree 20 to 60 ft. high with a dense crown of erect slender branches, or in the chaparral as a mere shrub; leaves oblong or oblong-lanceolate, entire, 3½ to 4½ in. long, on short petioles; umbels 4 to 9-flowered but setting only 1 to 3 (or rarely 6) fruits; flowers 2 lines long; drupe subglobose or ovoid, 1 in. long, greenish, or when ripe, brown-purple.—Cañon walls, mountain slopes and stream flats: Coast Ranges and Sierra Nevada, s. to San Diego Co. and n. to southern Ore. It is most abundant and of greatest size on the alluvial river flats of northwestern California and adjacent Oregon. Also called Bay Tree and Bay Laurel, but in the woods of Mendocino and Humboldt the name "Pepperwood" is the only one in use, while in Oregon the name "Myrtle" replaces all others The wood is prized by the cabinet-maker.

PLATANÀCEAE. Plane-tree Family

Large trees with alternate ample palmately lobed leaves and sheathing stipules; dilated base of petiole inclosing the bud of the next season; bark falling away in thin plates. Flowers in ball-like clusters scattered along a slender axis, the staminate and pistillate in separate axis. Fruit a coriaceous nutlet with tawny hairs about the base.

1. PLÁTANUS L. Plane-tree

The only genus. (Greek platus, broad, referring to the ample leaves.)

1. **P. racemòsa** Nutt. Western Sycamore. Tree 40 to 90 ft. high, typically

with huge wide-spreading limbs from a leaning, horizontal, or even prostrate trunk 1 to 5 ft. in diameter; leaves commonly broader than long, 4 to 12 (or 20) in. broad, parted into 3 to 5 broad spreading lobes; balls falling to pieces in winter, releasing the seed-like nutlets.—Stream bottoms, Great Valley, Sierra Nevada foothills and South Coast Ranges to coastal S. Cal.; also L. Cal.

ROSÁCEAE. Rose Family

Herbs, shrubs or trees with alternate leaves and persistent stipules. Flowers perfect, perigynous or epigynous. Calyx 5-lobed. Petals 5, rarely none. Stamens generally 10 to numerous, inserted with the petals on the margin of the disk lining the calyx-tube. Pistils 1 to many, distinct and free from the calyx, 1-celled with one style and stigma, or united into a 2 to 5-celled ovary, which is partly or completely inferior; styles as many as the carpels. Fruit a follicle, an achene, a drupe, a cluster of drupelets (as in a blackberry), or a pome.

Leaves alternate, simple.
  Ovary superior.
    Fruit an achene; flowers small, without petals... . 1. Cercocarpus
    Fruit a drupe; flowers with petals. . .. .2. Prunus
  Ovary inferior; fruit a pome; flowers with petals
    Flowers in a panicle; stamens 10; leaves evergreen . . 3 Heteromeles
    Flowers in a corymb; stamens 18 or 19; leaves deciduous .. ..4. Pyrus
Leaves opposite, pinnately divided, remarkably fern-like; ovary superior.. . .... .
4. Lyonothamnus

1. CERCOCÁRPUS H.B.K. Mountain Mahogany

Deciduous shrubs or low trees with spur-like branchlets and simple coriaceous straight-veined leaves. Flowers from winter buds, solitary or fascicled, terminal on the short branchlets. Calyx consisting of a slender pedicel-like tube abruptly expanded into the low-hemispherical deciduous 5-toothed limb. Petals none. Stamens numerous, borne in two or three rows on the calyx. Pistil 1, with a 1-celled ovary, 1 ovule and a single long style and terminal stigma. Fruit a villous achene inclosed in the persistent calyx-tube and surmounted by the very much elongated twisted soft-hairy style. (Greek kerkis, a shuttle, and karpos, fruit, in reference to the achene and its twisted tail.)

Flowers solitary or rarely in pairs; leaves narrowly lanceolate . ... .1. *C. ledifolius*
Flowers in 2 to 5-flowered clusters
  Leaves obovate, cuneate at base; clusters 2 or 3-flowered . . 2 *C. parvifolius*
  Leaves ovate to elliptic; clusters 4 or 5-flowered............ .. ... 3. *C. traskiae*

1. **C. ledifòlius** Nutt. Desert Mahogany. Shrub or scraggy tree, usually 6 to 20 or sometimes 40 ft. high, with a short trunk, ½ to 1 or rarely 2 ft. in diameter; leaves narrowly lanceolate, acute at both ends, entire with revolute margins, coriaceous, pale or rusty pubescent below, becoming glabrous and lustrous above, somewhat resinous, ½ to 1 in. long with a prominent midrib; flowers sessile, solitary or in pairs; calyx-limb deeply toothed; calyx-tube in fruit 4 or 5 lines long, the achene 3 lines long, and its tail 2 or 3 in. long.—Desert slopes of mts. bordering the Mohave Desert; e. slope of the Sierra Nevada; high desert ranges; inner North Coast Ranges from Lake Co. to Shasta Valley, thence e. to Modoc Co.; far e. to the Rocky Mts. Also called Mountain Mahogany.

2. **C. parvifòlius** Nutt. Hard Tack. Spreading shrub 5 to 8 ft. high; leaves obovate, serrate above the middle. cuneate and entire towards the base, not resinous, green above covered with a thin but dense felt below,

½ to 1½ in. long; flowers pedicelled, 2 or 3 in a cluster; calyx-limb shallowly toothed; calyx-tube in fruit 6 to 8 lines long, the achene nearly as long and the tail 1¼ to 4 in. long.—Common chaparral shrub throughout the Coast Ranges, Sierra Nevada and mts. of S. Cal.; e. to Rocky Mts. Often called Sweet Brush and also Mountain Mahogany, the wood very hard.

3. **C. tráskiae** Eastw. TRASK MAHOGANY. Shrub or small tree 10 to 20 ft. high; leaves broadly ovate to elliptic, dark green above, densely white woolly beneath, the parallel nerves fairly prominent, remotely crenulate above middle or seemingly entire by reason of the revolute margin, 1¼ to 2 in. long; flowers 4 or 5 in a cluster; achene 4 or 5 lines long, the tail 2 to 2¼ in. long.—Catalina Isl., s. side.

2. PRÙNUS L. PLUM

Shrubs or small trees. Leaves simple, serrate. Flowers white, in corymbs or in racemes from lateral buds borne on wood of the previous season, appearing before or with the leaves. Petals 5. Calyx 5-cleft, deciduous after flowering. Stamens 15 to 30. Pistil 1; style terminal. Drupe globose, without bloom; flesh sweet or bitter; stone globose or compressed, bony. (The Latin name of the Plum.)

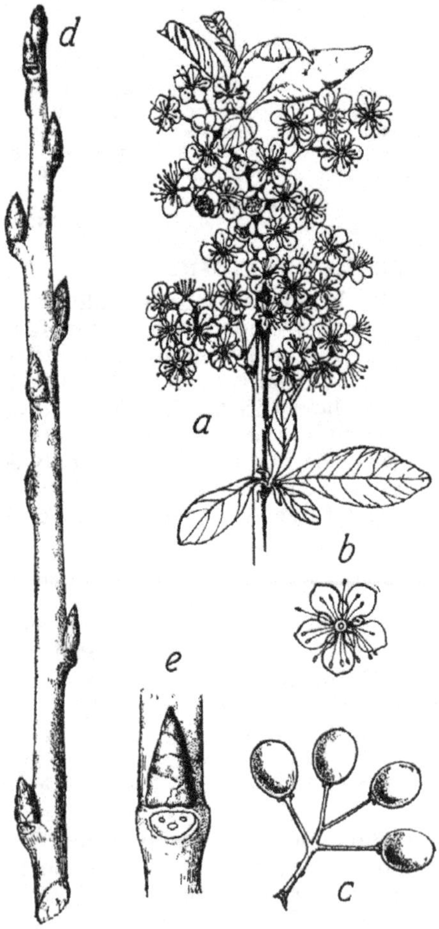

Fig. 118. Prunus emarginata Walp; *a*, fl branchlet x ½; *b*, fl x 1; *c*, fr. cluster x ½; *d*. winter branchlet x 1; *e*, leaf-scar and lateral winter bud x 2.

Leaves conduplicate in bud, drupe globose—CHERRIES
- Flowers in corymbs; drupe small, bright-red........ 1 *P. emarginata*
- Flowers in racemes; drupe dark-red
  - Peduncle leafy; drupe small, its flesh astringent; foliage deciduous.........2 *P. demissa.*
  - Peduncle leafless; drupe large, its flesh sweetish; foliage evergreen .. 3 *P. ilicifolia*

Leaves convolute in bud; flowers in umbels; drupe red, oblong—PLUMS .... 4 *P. subcordata*

1. **P. emarginàta** Walp. BITTER CHERRY. Fig. 118. Deciduous shrub 3 to 8 ft. high, rarely arboreous and 20 ft. high; bark smooth, dull red; leaves ovate or more commonly oblong-obovate, mostly obtuse, finely serrulate, ¾ to 1½ in. long, on petioles 1 to 3 lines long; flowers 3 to 10 in short corymbs; drupes 4 to 5 lines long, bright red, the pulp intensely bitter. — Forms extensive shrubby thickets on dry or moist gravelly slopes, or becoming tree-like near streams or on moist benches or flats, 4000 to 9000 ft., or near the sea from 500 to 1500 ft.: mts. of S. Cal.; Sierra Nevada; Coast Ranges; n. to Brit. Am.

2. **P. demíssa** Dietr. WESTERN CHOKE-CHERRY. Erect slender deciduous shrub 2 to 10 ft. high, or rarely a tree up to 22 ft. high; leaves oblong-ovate or more

commonly oblong-obovate, acute at apex or abruptly short-pointed, finely serrate, 1 to 3½ in. long; petiole ½ in. long; racemes 2 to 4 in. long, terminating more or less leafy peduncles; drupe red or dark purple, 3½ lines long, astringent.—Fertile mountain slopes or cañon flats, 2000 to 5000 ft., or near the sea ranging down to 500 ft.: higher ranges of S. Cal.; Sierra Nevada; Coast Ranges; n. to B. C., e. to the Rocky Mts.

3. **P. ilicifòlia** Walp. ISLAY. Evergreen shrub or small tree 5 to 28 ft. high; leaves coriaceous, elliptic or ovate, acute or obtuse, spinose-toothed, 1 to 2 in. long, short-petioled; racemes 1 to 2½ in. long, on axillary leafless peduncles; flowers 2 lines long; drupe red or dark purple, 6 to 8 lines thick, slightly obcompressed, apiculate; flesh thin, sweetish when ripe.—Mostly near the coast: Napa Range to the Santa Cruz Mts.; s. to coastal S. Cal. Var. INTEGRIFÒLIA Sudw. Island Cherry. Often a tree 15 to 45 ft. high; leaves oblong-ovate, usually entire, 3 to 6 in. long.—Santa Barbara Isls.

4. **P. subcordàta** Benth. SIERRA PLUM. Deciduous shrub 4 to 7 ft. high, or sometimes arborescent and 20 ft. high, with crooked and rough gray-brown branches and more or less spinescent branchlets; leaves ovate or elliptic to almost round, obtuse or truncate at base, rarely subcordate, 1 to 2 in. long, on petioles 2 or 3 lines long; flowers appearing with the leaves, 2 to 4 in a cluster, on pedicels ½ in. long; drupe red, ¾ to nearly 1 in. long, the pulp rather hard but more or less edible.—Sierra Nevada from Tulare Co. to Siskiyou Co., thence s. in the Coast Ranges to the Santa Cruz Mts.

3. HETEROMÈLES Roem.

Evergreen shrub with simple coriaceous serrate leaves. Flowers white, small, numerous, in little cymes disposed in a terminal corymbose panicle. Calyx turbinate, 5-cleft. Petals spreading. Stamens 10, in pairs opposite the calyx-teeth; filaments dilated at base and somewhat connate. Pistils 2 or 3, lightly united, tomentose above, only lightly adherent to the fleshy calyx-tube, the thickened persistent calyx-teeth closed over them in fruit. Fruit bright red, ovoid, berry-like. Seeds 1 or 2 in each cell. (Greek heteros, different, and melon, an apple.)

Fig 119 Heteromeles arbutifolia Roem ; fr. branchlet x ¼.

1. **H. arbutifòlia** (Lindl.) Roem. CHRISTMAS BERRY. TOYON. Fig. 119. Shrub, rarely a small tree, 5 to 15 ft. high; leaves oblong, acute at base and apex, dark green, lighter beneath, sharply serrate, 2 to 4 in. long, on petioles ½ to ¾ in. long; panicle in anthesis rather dense, 2 to 3 in. high; corolla 2½ lines in diameter; fruit 3 or 4 lines long, the seeds obovate, flat on one side, convex on the other, ½ as long.—Stony slopes, stream banks and gulches in the foothills: Sierra Nevada; Coast Ranges; s. to S. Cal. and L. Cal. Abundant from Napa Co. to Mendocino Co. One of the most handsome of Californian shrubs when covered from Nov. to Jan. with its fine clusters of crimson berries.

4. PŶRUS L. Pear. Apple

Trees or shrubs with simple deciduous leaves and stipules which disappear early. Flowers in corymbs. Calyx-tube urn-shaped. Petals white or pink, with claws. Styles 2 to 5, united at base; ovules 2 in each cell of the inferior ovary, the carpels more or less coriaceous. Fruit a pome, in the subgenus Malus (apple) more or less globose and sunken at each end. (Latin name of the pear.)

1. **P. diversifòlia** Bong. Oregon Crab-apple. Tree 15 to 30 ft. high, or often a many-stemmed shrub; leaves ovate, pointed, serrate, green above, pale, pubescent and eventually rusty beneath, 1 to 3¾ in. long, those of the sterile branchlets mostly 3-lobed or with a coarse tooth on each side, those of the flowering branchlets rarely lobed or toothed; corymbs 6 to 8-flowered; petals elliptical, 5 lines long; stamens 18 or 19; fruits 2 or 3 in a cluster, oblong or oblong-ovoid, 6 to 7 lines long and 4½ or 5 lines broad, not sunken at base, yellowish (or pinkish on one side), aging purple-black; calyx-lobes at length deciduous.—North Coast Ranges, mostly near the coast, Sonoma to Eureka; n. to Wash. and Alas. Rare with us. (P. rivularis Dougl.)

5. LYONOTHÁMNUS Gray

Evergreen shrub or tree with thin bark exfoliating in long loose strips and opposite dimorphic petioled leaves. Flowers numerous in a much-branched terminal panicle. Petals 5 and stamens 13 to 16, inserted on the margin of the woolly disk lining the calyx-tube which bears 5 lobes. Pistils 2, distinct, each with a spreading style and capitate stigma. Fruit consisting of two woody 4-seeded dehiscent carpels. (W. S. Lyon, the discoverer, his surname in combination with Greek thamnos, shrub.)

1. **L. floribúndus** Gray. Catalina Ironwood. Slender tree 15 to 55 ft. high with narrow crown and often tall trunk 3 to 12 in. in diameter; leaves oleander-like, linear, nearly entire or pinnately cut, petioled, 3 to 5 in. long, or often pinnately compound with 2 to 5 leaflets similar in shape and size to the simple leaves; flowers white, 3 lines broad, in terminal clusters 3 to 6 in. broad.—Insular species confined to Santa Catalina, San Clemente, Santa Rosa and Santa Cruz isls.

LEGUMINÒSAE. Pea Family

Herbs, shrubs, or trees. Leaves alternate, stipulate, compound or rarely simple. Flowers perfect, somewhat perigynous (frequently more on one side than the other), or hypogynous. Calyx synsepalous, usually 5-toothed or cleft, or sometimes bilabiate, mostly persistent. Corolla with 5 petals, regular or in ours commonly papilionaceous, i. e., highly irregular and butterfly-like, with an upper petal or "banner," 2 lateral petals or "wings," and the 2 lower petals joined by their edges to form the "keel," usually with free claws. Stamens 10, united into a sheath around the ovary (monadelphous), or the upper stamen distinct from the others (diadelphous), or sometimes all distinct. Pistil 1, 1-celled, with single style and stigma. Fruit a 2-valved pod or legume, with 1 or 2 rows of seeds on the ventral side, commonly opening by both the dorsal and ventral sutures, or sometimes indehiscent. Seeds without endosperm.

Stamens distinct; corolla regular or imperfectly papilionaceous; leaves bipinnate; branches more or less spiny

Flowers small, regular; calyx campanulate; stamens much exserted.....1. Prosopis

Flowers medium-sized, the upper petal larger; calyx with stipe-like tube; stamens included.....2. Cercidium.

Stamens monadelphous or diadelphous; corolla papillionaceous.
Leaves simple, glandular-dotted; branchlets numerous, spinose........3. PAROSELA
Leaves once pinnate, without glandular dots; spines in pairs below petioles 4 OLNEYA,

1. PROSÒPIS L.

Deciduous shrubs or trees, the branches armed with spines and without terminal buds. Leaves alternate on the season's shoot, fascicled in earlier axils, deciduous, bipinnate with 1 or 2 pairs of pinnae, the leaflets small, numerous, entire and in equal pairs. Flowers small, greenish, regular, sessile, in axillary pedunculate cylindrical spikes. Calyx campanulate, with short teeth, deciduous. Petals 5, connate below the middle or at length free, tomentose on inner side, very much exceeding the calyx. Stamens 10, free, exserted, the anthers tipped with a deciduous gland. Ovary stipitate, villous; style filiform. Pod straight, curved or coiled, indehiscent, the many seeds separated by thick spongy partitions. (Greek prosopis, ancient name for the butter-bur.)

Leaflets 10 to 15 pairs; spines axillary, in pairs, singly or none; pod straight or curving, compressed...1. *P. juliflora*
Leaflets 5 to 8 pairs; spines in pairs, stipular; pod spirally coiled into a straight cylindric body ....2 *P. pubescens*

1. **P. julifòra** DC. HONEY MESQUITE. Fig. 120. Tree or shrub, the short trunk dividing into wide-spreading branches forming a round or low broad crown commonly 10 to 20 ft. high and usually twice or thrice as broad; petioles abruptly enlarged and glandular at base, bearing usually 2 pinnae, with 10 to 15 pairs of linear leaflets ½ to 1 in. long; stipules linear and membranaceous; spines axillary, often in pairs or sometimes absent, ¼ to 1¼ in. long; flowers 2 lines long, condensed in slender cylindrical spikes mostly 2 to 3½ in. long; stamens twice as long as the petals; pods borne in drooping clusters, 1 to 6 to each spike, linear, at first flat, later becoming thickened, falcate, 4 to 8 in. long, irregularly constricted between the seeds which are about 3 lines long.—Colorado and Mohave deserts; local in the upper San Joaquin Valley, at San Jacinto Lake and in w. San Diego Co.; e. to Tex., s. to northern Mex.

Fig. 120. Prosopis juliflora DC.; fl branchlet x ⅓.

2. **P. pubéscens** Benth. SCREW-BEAN MESQUITE. Fig. 121. Shrub or small tree, 10 to 25 ft. high with trunk 3 to 10 in. in diameter and branches armed with stout stipular spines 2 to 6 lines long; leaves canescently puberulent, the leaflets in 5 to 8 pairs, oblong, 1 to 5 lines long; flowers 2 lines long, borne in spikes 2 to 3 in. long, each spike setting 2 to 15 pods; pod coiled into a narrow straight cylindric body 1 to 1½ in. long; seeds less than 1 line long.—Sandy or gravelly washes or ravines: Colorado and Mohave deserts; n. to Death Valley, e. to N. Mex., s. to northern Mex. The beans are sweet and nutritious and are used as food by the Indians and fed as fodder to cattle.

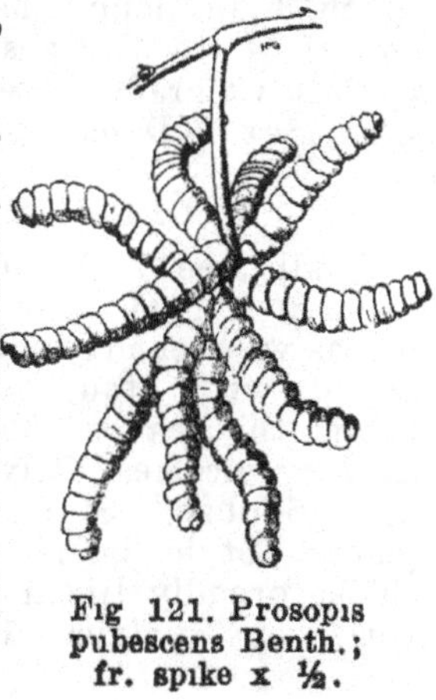

Fig 121. Prosopis pubescens Benth.; fr. spike x ½.

2. CERCÍDIUM Tul.

Shrubs or small trees, often armed with short spines. Leaves bipinnate with one or two pairs of pinnae and 2 to 4 equal pairs of leaflets. Flowers on jointed pedicels in short axillary racemes. Calyx shortly campanulate, the limb cleft into 5 reflexed deciduous lobes. Petals bright yellow, clawed, the upper one broader than the rest, a little auricled at base of blade, and with longer claw. Stamens 10, distinct, the filaments hairy at base, one or two next upper petal gibbous on one side toward base. Pod compressed, 2-valved, narrow, pointed at each end, more or less constricted between the flat seeds. (Greek cercidion, a weaver's shuttle, in reference to the fruit.)

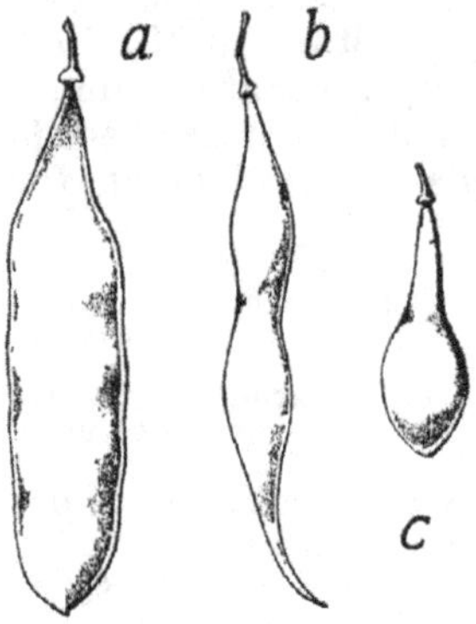

Fig. 122. Cercidium torreyanum Sarg.; *a, b, c,* different forms of pods x ½.

1. **C. torreyànum** (Wats.) Sarg. PALO VERDE. Fig. 122. Small tree 15 to 20 ft. high, with erect branches, short trunk, smooth green bark, and leafless for most of the year; spines 2 or 3 lines long; leaflets oblong, 2 to 4 lines long; flowers ¾ in. broad; petals orbicular to ovate, 4 to 5 lines long; pods 2 to 3 in. long, with a double groove along the ventral suture, often conspicuously constricted between the seeds.—Sandy washes, Colorado Desert; e. to Ariz., s. to Mex.

3. PARÒSELA Cav.

Glandular punctate herbs, small shrubs or small trees. Leaves unequally pinnate or simple; stipules small, subulate; leaflets small, entire. Flowers in terminal spikes or racemes. Calyx with 5 nearly equal teeth or lobes, persistent. Banner cordate with free claw; claws of wings and keel adnate to the cleft staminate tube. Stamens 10, rarely 9, monadelphous. Pod ovate, compressed, usually indehiscent, more or less included in the calyx, 1 to 2-seeded. Seeds reniform. (Anagram of Psoralea.)

1. **P. spinòsa** (Gray) Hel. SMOKE BUSH. Fig. 123. Very spinose and nearly leafless ashy-gray shrub or small tree 4 to 25 ft. high with intricately much-branched top; leaves few, simple, soon deciduous; spikes ½ to 1 in. long, the violet-purple flowers 4 to 5 lines long.—Sandy washes, Colorado Desert; e. to Ariz., s. to Mex. (Dalea spinosa Gray.)

Fig 123. Parosela spinosa Hel.; *a,* fl. branchlet x ¾; *b,* fl x 2; *c,* banner x 2; *d,* pod x 2.

4. ÓLNEYA Gray

Small tree with thin scaly bark, slightly angled branchlets, pinnate leaves with entire leaflets, and often armed with stout spines in pairs below the leaves. Flowers few, in axillary racemes. Calyx subcampanulate, 5-lobed. Stamens 10, diadelphous. Style bearded above. Pod thick, broadly linear or ovate, with coriaceous valves, 1 to 8-seeded, tardily dehiscent. (Stephen T. Olney, Rhode Island botanist of the 19th century.)

1. **O. tesòta** Gray. DESERT IRONWOOD. A spreading tree 15 to 20 ft. high with short trunk ¼ to 1 (or 1½) ft. in diameter; leaflets 5 to 7 pairs, cuneate-oblong or -obovate, obtuse or notched at apex, 3 to 6 lines long; stipular spines stout, straight or slightly curved, 1 to 4 lines long; flowers violet-purple, 4 or 5 lines long, 3 to 10 in loose racemes ½ to 1¼ in. long; pods glandular, more or less pubescent and often provided with tack-shaped glands, ½ to 3 in. long, more or less constricted between the seeds.—Desert valleys, Colorado Desert; e. to Ariz., s. to Mex. Its wood is remarkably hard and heavy and is used by desert Indians for arrow parts and tool-handles. The leaves persist over winter until spring and fall before the new leaves appear.

SAPINDÀCEAE. BUCKEYE FAMILY

Deciduous trees or shrubs with opposite compound leaves, no stipules, and slightly irregular flowers. Ovary superior, 3-celled with 2 ovules in each cell, commonly but one ovule maturing. Endosperm none.

1. AÉSCULUS L. HORSE CHESTNUT

Leaves palmately compound with serrate leaflets. Flowers showy, ill-scented, on jointed pedicels in a terminal cylindrical thyrse, of two sorts, perfect (fertile) with long thick styles and sterile with short styles; fertile flowers few near top of thyrse. Calyx tubular, unequally cleft. Petals 4 or 5, slightly unequal, clawed. Stamens 5 to 7, becoming successively much exserted and often unequal. Fruit a large 3-valved capsule releasing one large polished seed. (Latin name of an Italian oak with edible acorns.)

1. **A. califòrnica** (Spach) Nutt. BUCKEYE. A low tree (commonly 10 to 15 ft. high) with a rounded or depressed crown of greater breadth; leaflets 5 to 7, oblong-lanceolate to oblong-elliptic, acute or acuminate, 3 to 5 in. long; thyrse 4 to 6 in. long; petals 6 or 7 lines long, the elliptic or ovate limb rotately spreading; axis of the thyrse at length naked and pendulous, bearing one pear-like pod or sometimes 2 to 9; seed 1 to 2 in. in diameter.—Low dry hills or cañon sides or moist valley flats, Sierra Nevada foothills and Coast Ranges from Shasta Co. to the Tehachapi Mts., thence s. a short distance to Sawmill Mt. Fl. June.

ACERÀCEAE. MAPLE FAMILY

Trees or shrubs with opposite leaves. Flowers regular, polygamous or dioecious, borne in racemes, corymbs or fascicles. Calyx generally cleft into 5 segments, the petals as many or none. Stamens 3 to 10, borne on the edge of a disk or hypogynous. Pistil 1 with a 2-lobed 2-celled ovary and 2 styles. Ovary developing a long wing from the summit of each lobe and thus ripening into a double samara; samaras separable at maturity, the wings serving to rotate them rapidly in the air and further their horizontal flight when carried away by the wind.

1. ÀCER L. MAPLE

Leaves simple and palmately lobed, or pinnately compound, always petioled. Flowers small, the clusters always drooping. (Latin name of the Maple tree.)

Leaves simple; petals present
  Flowers in racemes, samaras more or less hispid; leaves large, deeply 5-lobed . . . 1. *A. macrophyllum*
  Flowers in corymbs; samaras glabrous
    Leaves shallowly but acutely 7 to 9-lobed; sepals nearly twice as long as the petals . . . . 2 *A circinatum.*

Leaves mostly 3-lobed or parted; sepals equaling petals..........3. *A. glabrum*
Leaves pinnately or ternately compound; flowers dioecious; petals none....4 *A. negundo*

1. **A. macrophýllum** Pursh. BIG-LEAF MAPLE. Broad-crowned tree 30 to 95 ft. high; leaves roundish in outline, palmately parted into 5 broad fingers, 4 to 10 in. broad; perfect and staminate flowers mixed in the same raceme; stamens 7 to 9, villous below; body of samara short-bristly, the wings 1 to 1½ in. long.—Along living streams throughout cismontane California and also on east and north slopes of mt. ranges in the areas of about 25 to 50 inches rainfall; n. to B. C. and Alas. Also called Water Maple, White Maple and Oregon Maple.

2. **A. circinàtum** Pursh. VINE MAPLE. Shrub or sometimes a small tree, erect and 5 to 20 ft. high, but more often vine-like or reclining; trunk 3 to 6 in. or rarely 1 ft. in diameter; leaves 2 to 4 in. broad, 5 to 7-lobed to the middle, with toothed margin; flowers 4 to 10 or more in a corymb, most of them staminate, the cluster often setting but one fruit; stamens 6 to 10, shorter than the petals in the perfect flower but longer than the petals in the staminate flower; filaments villous below; samaras glabrous, the wings spreading at right angles to the stalk; wings 7 to 10 lines long, 4 to 5 lines wide, scarlet when full grown; as the fruit ripens, the peduncle turns upward and finally the samara stands erect above the leaf.—Humboldt Co.; n. to B. C.

3. **A. glàbrum** Torr. SIERRA MAPLE. Shrub 5 to 10 ft. high with slender branchlets, the trunk 2 or 3 in. in diameter; leaves 1 to 3 in. broad, palmately 3-lobed or often with 2 supplementary lobes at base, the margin unequally serrate; flowers 4 to 9, in loose umbel-like corymbs, the staminate without rudiments of pistils and the pistillate with short stamens; cormybs unisexual or with both pistillate and staminate flowers, the sexes often borne on different shrubs; stamens 7 to 10, the glabrous filaments arising from pits in the perigynous cushion; samaras usually several in a cluster, glabrous, with diverging wings 6 to 12 lines long and 4 or 5 lines wide.—Rocky or wet mountain sides, 6000 to 8000 ft.: San Jacinto Mts.; San Bernardino Mts.; Sierra Nevada; Salmon Mts. to Siskiyou Mts.; n. to Alas., e. to the Rocky Mts.

4. **A. negúndo** L. var. **califòrnicum** Sarg. BOX ELDER. Tree 20 to 60 ft high; leaves pinnately 3-foliolate, the leaflets 1¼ to 5 in. long, serrate and incised, or deeply 2 or 3-lobed, or the lobes sometimes becoming distinct and petioled so that one or more of the primary leaflets is replaced by 2 or 3; staminate flowers clustered on thread-like hairy pedicels, the stamens 4 or 5; pistillate flowers borne in slender racemes; samaras straw-white, crimson when young, finely pubescent, the wings 6 to 8 lines long.—Along streams and in low moist valley bottoms: Sierra Nevada foothills; Coast Ranges; coastal S. Cal.

CACTÀCEAE. CACTUS FAMILY

Trees or shrubs with fleshy or woody stems more or less studded with clusters of thorns (modified leaves), and without normal foliage. Flowers complete. Sepals and petals many, passing one into the other. Stamens numerous, inserted on the base of the corolla. Pistil compound; ovary inferior, 1-celled, with many parietal placentae; style one with many branches. Fruit fleshy, berry-like; seeds without endosperm.

CÉREUS Haw.

Habit diverse. Cushions geminate, borne on the vertical ribs, the lower spine-bearing, the upper producing a branch or flower. Calyx elongated. Stamens adnate at base to tube of calyx. Seeds black.

1. **C. gigantèus** Engelm. SUWARRO. Columnar tree, strongly 8 to 12-ribbed or fluted toward the base and with more numerous ribs above, 15 to 60 ft. high, simple or with 2 or 3 or more upright branches above the middle; foliage leaves none; flowers 4 to 4½ in. long; fruit oval, dehiscent by irregular valves.—Riverside Mts. along the Colorado River; e. to Ariz. Its singular columnar growth and candelabra-like branches make it an extraordinary feature in the desert landscapes of the regions it inhabits.

CORNÀCEAE. Dogwood Family

Ours deciduous trees or shrubs. Leaves opposite, simple, entire. Flowers in cymes or heads. Calyx-tube coherent with the ovary, its limb represented by 4 small teeth at the summit or none. Petals 4, distinct, epigynous, valvate in bud. Stamens 4, alternate with the petals. Ovary 2-celled with a single pendulous ovule in each cell; style 1, filiform.

1. CÓRNUS L. Cornel. Dogwood

Flowers regular and perfect, greenish or white, disposed in cymes or heads. Fruit a drupe, the stone 2-celled with 1 seed in each cell. (Latin cornu, horn, on account of the hardness of the wood.)

1. **C. nuttállii** Aud. MOUNTAIN DOGWOOD. Small tree 10 to 30 ft. high; leaves narrow- or elliptic-obovate or even orbicular, with rounded or shortly acute apex, 3 to 5 in. long, on petioles 2 to 3 lines long; flowers crowded in a head on a thick convex receptacle and surrounded by a showy petal-like involucre; bracts of the involucre commonly 6, white, sometimes tinged with red, obovate to oblong, abruptly acute or acuminate, 1½ to 3 in. long; drupe 5 to 6 lines long, scarlet.—Depth of woods in deep moist soils in the mts., 2500 to 5100 ft.: Palomar Mt.; San Jacinto Mts.; San Bernardino Mts.; Sierra Nevada; North Coast Ranges; n. to B. C.

ERICÀCEAE. Heath Family

Trees, shrubs or perennial herbs. Leaves simple, commonly alternate. Flowers regular, the parts usually in 5s. Stamens free or almost free from the corolla, as many or twice as many as its lobes; anthers 2-celled, opening by a terminal pore. Ovary superior or inferior, commonly 4 to 10-celled, with axile placentae and numerous ovules. Fruit a capsule or indehiscent and either dry or fleshy.

1. ÁRBUTUS L. Arbute Tree

Evergreen trees or shrubs with glossy leathery leaves. Flowers white, in a terminal panicle of dense racemes. Calyx small, 5-parted. Corolla globular or ovate, 5-lobed at apex. Stamens twice as many as the corolla-lobes, included; filaments soft hairy; anthers with a pair of reflexed awns on the back, each cell opening at the apex anteriorly by a pore. Ovary on a hypogynous disk, 5 or rarely 4-celled, the ovules crowded on a fleshy placenta which projects from the inner angle of each cell. Fruit a many-seeded berry with granular surface. (Latin name of the Arbute tree, under which, says Horace, idle men delight to lie.)

1. **A. menzièsii** Pursh. MADROÑO. Widely branching tree 20 to 125 ft. high; bark polished, crimson or terra cotta, or on old trunks dark brown and fissured into small scales; leaves narrowly elliptic or ovatish, 3 to 6 in. long, glabrous, dark green and polished above, glaucous beneath, entire, or on vigorous shoots, finely serrate; corolla 3 lines long, with 5 very small lobes recurving from the small opening, and 10 semitransparent glands in a circle at base with a slight constriction above them which becomes obvious on

drying; berry somewhat depressed globose, 4 to 5 lines in diameter, fleshy but rather dry, red or orange color; seeds somewhat angular, closely crowded, 5 or 6 in a cell.—Foothills and mountains, 300 to 4000 ft.: North Coast Ranges (where it reaches its greatest development); South Coast Ranges; local in mts. of S. Cal. (Santa Ynez, San Gabriel, San Bernardino and Palomar mountains); Sierra Nevada from Tuolumne Co. to Shasta Co.; n. to B. C.

OLEÀCEAE. Ash Family

Trees or shrubs mostly with opposite leaves. Flowers small, commonly in panicles, mostly unisexual. Stamens few. Ovary superior, 2-celled; style one. Fruit a samara, capsule or drupe.

1. FRÁXINUS L. Ash

Trees or shrubs. Leaves pinnately compound (except one species), deciduous. Flowers in small panicles, appearing just before the leaves and from separate buds. Corolla with 2 equal petals or in our trees none. Stamens 2 (rarely 1 or 3). Ovules 2 in each cell. Fruit a 1-seeded samara, with terminal wing. (Latin name of the ash.)

Trees; corolla none; style conspicuously 2-lobed
- Flowers dioecious, leaves pinnate; leaflets 2 in long or more; branchlets terete.
  - Leaflets oblong to oval, the lateral commonly sessile; cismontane...1 *F oregona*
  - Leaflets lanceolate to ovate, the lateral leaflets on petioles ⅛ to ½ in. long; desert ranges ....2 *F. velutina*
- Flowers polygamus, leaves simple, rarely with 2 or 3 leaflets; branchlets of the season 4-sided; desert area... .... 3. *F. anomala*

Shrubs; corolla present; flowers perfect; style obscurely lobed; leaflets mostly stalked, less than 2 in long, branchlets of the season strongly 4-sided; cismontane 4 *F dipetala*

1. **F. óregòna** Nutt. Oregon Ash. Fig. 124. Tree 30 to 80 ft. high; leaflets 5 to 7, oblong to oval, or often broadest toward the apex and abruptly short-pointed; 2 to 5½ in. long; flowers borne in small crowded clusters; samara oblong-lanceolate, 1¼ to 2 in. long.—Along streams and in valley or lake bottoms, throughout cismontane California, especially common in the North Coast Ranges; n. to B. C.

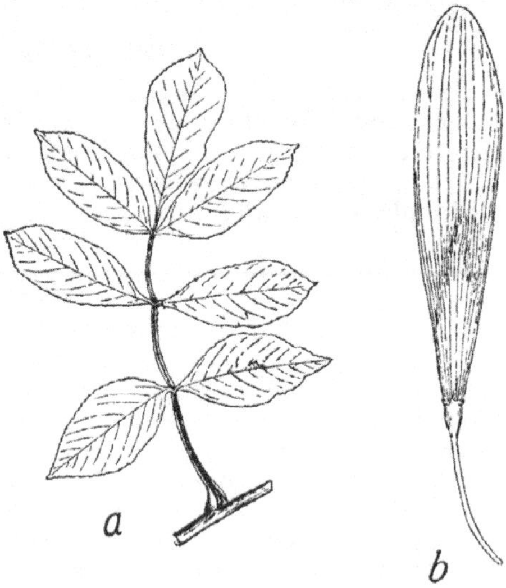

Fig. 124. Fraxinus oregona Nutt , *a*, leaf x ¼ ; *b*, fr x 1

2. **F. velùtina** Torr. Arizona Ash. Tree 15 to 30 ft. high; leaflets 5, lanceolate to narrowly ovate, less acute or less commonly attenuate, 2 to 4 in. long; samaras 1 in. long, the wing 1½ to 2 lines wide.—Along desert streams or the borders of lakes or springs: Owens Lake; Panamint Range; sw. base Mt. San Jacinto; betw. Campo and Jacumba; e. to Tex. The leaflets, while sometimes very willow-like, are also ovate and there are intergrades to the var. coriàcea Jepson n. comb. Leather-leaf Ash. Leaflets narrowly or broadly ovate, more or less abruptly acute or attenuate.—Range of the species in Cal., thence e. to Utah. (F. coriacea Wats.)

3. **F. anòmala** Wats. Dwarf Ash. Tree 15 to 20 ft. high, or a low spreading shrub; leaves simple, roundish or broadly ovate, 1½ to 2 in. long, or

compound with 2 or 3 similar leaflets; flowers either perfect or pistillate, both forms occurring in the same cluster; samara 8 or 9 lines long, with a rounded wing which surrounds the body and is 4 to 5 lines broad.—Desert ranges, Panamint and Providence mountains; e. to Col. and Ariz.

4. **F. dipétala** H. & A. MOUNTAIN ASH. Shrub 5 to 15 ft. high; leaves 2 to 6 in. long; leaflets 3 to 9, serrate above the middle, ¾ to 1½ in. long; petals 2, white, about 3 lines long; samaras 1 to 1¼ in. long, the wing frequently notched at tip.—Cañons and mt. slopes, 1500 to 4000 ft.: Sierra Nevada; Coast Ranges; coastal S. Cal.

BIGNONIÀCEAE. BIGNONIA FAMILY

Trees or shrubs, the leaves most commonly opposite, in ours simple. Flowers large and showy, perfect, 2-lipped. Stamens 4 in 2 pairs, the fifth stamen sterile or wanting. Ovary 2-celled, style 1; stigma 2-lobed. Valves of the fruit 2, falling away from the placentiferous partition and releasing usually winged seeds.

1. CHILÓPSIS Don.

Flowers in a short terminal raceme. Corolla funnelform, ventricose above, the limb ample. Stamens 4 and a sterile filament. Capsule long, linear, terete. Seeds oblong, thin, the wing at each end replaced by a tuft of soft hairs. (Greek cheilos, lip, and opsis, resemblance.)

1. **C. salígna** Don. DESERT WILLOW. Deciduous shrub or tree with either broad or narrow crown, 10 to 25 ft. high; leaves opposite whorled or mostly irregularly alternate linear with lanceolate apex, entire, 1½ to 3 lines wide, 2 to 5 in. long; corolla white and pink-lavender, blotched with yellow in throat, 1 to 1¼ in. long, its rounded spreading lobes erose and undulate; capsule 6 to 12 in. long and 2 lines broad, with oblong thin seeds 4 lines long.—Washes, Mohave and Colorado deserts, recurring locally in San Jacinto Valley; e. to southern Nev. and western Tex., s. to Mex.

CAPRIFÒLIACEAE. HONÈYSUCKLE FAMILY

Small trees or shrubs with opposite leaves. Flowers complete. Calyx-tube adnate to the ovary, the toothed limb insignificant. Corolla tubular or rotate, regular or irregular, 4 or 5-lobed. Stamens inserted on the corolla, as many as its lobes. Ovary 2 to 5-celled; style one. Fruit in ours a berry or berry-like drupe.

1. SAMBÙCUS L. ELDERBERRY

Deciduous shrubs or small trees. Leaves compound, odd-pinnate with serrate leaflets. Flowers small, white, in a terminal compound cyme, jointed with their pedicels. Calyx 5-toothed. Corolla regular, rotate, deeply 5-lobed. Ovary 3 to 5-celled; style short; stigmas 3 to 5; ovules solitary, suspended from the summit of each cell. Fruit a small globose berry-like drupe containing 3 to 5 cartilaginous nutlets. (Greek sambuke, a musical instrument, said to have been made of elder wood.)

Cymes flat-topped; berry blue with a bloom; winter buds small
  Leaves glabrous, the leaflets equal-sided at base . . . . . 1. *S. glauca*
  Leaves finely pubescent beneath, the leaflets unequal-sided at base . 2. *S. mexicana.*
Cymes dome-shaped or thyrsoid; berry red; winter buds large with large scales. . . . . . 3. *S. racemosa*

1. **S. glaùca** Nutt. BLUE ELDER. Roughish thickety bush 5 to 10 ft. high or eventually developing a small tree with distinct trunk, up to 25 ft. high; leaves glabrous; leaflets 5 to 7, ovate to oblong-lanceolate, serrate except at the abruptly acuminate apex, 1 to 3½ in. long; cymes flat-topped, 2 to 6 or 9

in. broad; flowers white, 2½ to 3½ lines broad; berries 2 lines in diameter, blue beneath the white bloom.—Open woods, cañons or moist flats of the lower hill country or middle altitudes, or along stream-banks in the valleys· Coast Ranges; Sacramento and San Joaquin valleys; Sierra Nevada; n. to B. C., e. to Utah. Fl. May-Aug., fr. Aug.-Oct.

2. **S. mexicàna** DC. Shrub or small tree 6 to 8 or 15 ft. high; leaves thick or coriaceous, finely and often densely short-pubescent, especially beneath, or sometimes glabrous; leaflets 5 to 7 or 9, thickish or coriaceous, oblong-ovate, abruptly attenuate at apex, finely serrate, one side continued farther down the petiolule than the other and so very unequal at base, 3 to 6 (or 8) in. long, the lower leaflets not infrequently with a supplementary leaflet; cymes flat-topped, 4 to 8 or 16 in. broad; flowers yellowish; berries black with a bloom. —Dry open hillsides, 3000 to 7000 ft.: Sierra Nevada; mts. of S. Cal.; e. to Tex., s. to Mex. Aug.

3. **S. racemòsa** L. Thick spreading bush 1 to 3 (or 4) ft. high; leaves thin, glabrous or nearly so; leaflets 5, ovate to elliptic, equal-sided at base, narrowed at apex to a slender entire point, serrate, 1½ to 3 (or 5) in. long; cymes round-ovate, 1¼ to 3 in. high; flowers cream-color; berries red, 2½ lines in diameter.—Moist places and especially the margins of surface streams on mountain slopes: Sierra Nevada, 7000 to 10,500 ft.; n. to B. C., e. to the Atlantic. Eur. July. Var. CALLICÁRPA Jepson. Low shrub or small tree, 8 to 14 or 20 ft. high; leaves thinnish, pubescent beneath and often above with short appressed hairs; leaflets thinnish, oblong-ovate or obovate, acuminate, sharply serrate to the very apex, 2 to 7 in. long; berries scarlet, without bloom, 2 lines broad; fruiting clusters 2 to 5 in. across, very showy. —Cañon beds and flats near the sea from San Mateo Co. to Humboldt Co.; n. to Wash.

GEOGRAPHICAL INDEX

Coast Ranges, the complex of ranges west of the Great Valley.

Great Valley, the great central valley of California, including the Sacramento and San Joaquin valleys.

Inner Coast Range, the range next to the Great Valley.

Mt. Hamilton Range, the range east of Santa Clara Valley from Livermore Valley south to Pacheco Pass.

North Coast Ranges, the ranges north of San Francisco Bay as far as the Siskiyous.

San Bernardino Mountains, the range from Cajon Pass easterly to Morongo Pass.

San Carlos Range, the inner South Coast Range from Panoche Pass south to Warthan Creek.

San Gabriel Mountains, the range from Soledad Cañon easterly to Cajon Pass.

Santa Barbara Islands, collective name for all the islands off the coast of Southern California.

Santa Cruz Mountains, the range from Pajaro River north through San Mateo County to the San Bruno Hills.

Seaward Coast Range, the range next the ocean; applied more especially to the seaward North Coast Range.

Sierra Nevada, the main mountain axis on the eastern side of the Great Valley, south to Tehachapi Pass and north to Pitt River.

South Coast Ranges, the ranges south of San Francisco Bay as far as Santa Barbara County (Santa Maria River).

Yollo Bolly Range, the inner North Coast Range from Snow Mountain north to North Yollo Bolly.

GLOSSARY

Achene, a dry 1-seeded indehiscent fruit; strictly, one derived from a simple pistil.

Acorn, the fruit of an oak, consisting of nut and cup; cf. Figs. 47, 48, 49.

Acute, ending in an acute angle but not tapering or prolonged.

Adherent, same as adnate.

Adnate, said of different organs or parts which are grown together from the first.

Blade, the expanded portion of a leaf or petal.

Bract, a modified or undeveloped leaf or scale-like organ subtending a flower or a flower-branch; a usually narrow and often minute structure subtending the cone-scale in coniferae (Figs. 34, 37, 38).

Bur, a spiny fruit like that of a chinquapin or chestnut.

Caducous, dropping or falling early, especially in advance of other parts; falling easily.

Calyx, the outer usually green envelope of the flower.

Capitate, head-like (Fig. 83c).

Capsule, a dehiscent fruit derived from a compound pistil.

Carpel, a simple pistil or one of the parts of a compound pistil.

Catkin, a densely flowered scaly spike which falls whole after flowering or after maturity (Fig. 61); also applied to the flowering cones in coniferae.

Choripetalous, with distinct petals.

Clavate, narrow and tapering gradually from base to apex; club-shaped.

Claw, the narrow or stalk-like base of a petal or sepal.

Coherent, same as Connate.

Complete flower, with all four circles.

Compound leaf, with the blade completely divided into several distinct parts or leaflets (Figs. 85 and 124).

Connate, said of similar parts more or less grown together.

Conduplicate, with the two sides or halves (as of a leaf) placed face to face.

Convolute, rolled up from one edge to the other.

Coriaceous, leathery.

Corolla, the inner usually colored envelope of the flower.

Corymb, a flat-topped flower-cluster, the outer flowers with longest pedicels and blooming first.

Crenate, with rounded teeth.

Crenulate, finely crenate.

Cuneate, wedge-shaped.

Cyme, a flat-topped flower-cluster, the central flowers opening first (compound cyme, Fig. 85).

Deciduous, barren of leaves in winter; falling after having performed its function (said of corollas and similar parts).

Decurrent, running down, as the blade extending down the petiole or on to the stem.

Deflexed, turned abruptly downward.

Dehiscent, splitting open.

Dentate, toothed with the teeth pointing straight out from the margin (Fig. 53).

Dilated, expanded or flattened, like the blade of a leaf.

Dimorphic, of two forms; compare dimorphic foliage of Redwood, Figs. 4 and 5.

Dioecious, with stamens and pistils in separate flowers on different plants.

Dorsal, the side turned away from the axis of growth; lower; inferior; back. cf. *ventral.*

Drupe, a fruit with two layers about the seed, the inner hard and stony, the outer fleshy.

Elliptical, a little longer than broad, and with curving margin; like an ellipse.

Emarginate, notched at apex.

Embryo, the plantlet in the seed.

Endosperm, the reserve tissue of the seed in which the embryo is usually embedded.

Entire, the margin whole and even, not toothed or lobed.

Epigynous, with corolla and stamens borne on the summit of the ovary or seemingly so.

Extrorse, situated on the outside or directed outwards.

Fascicled, in a bundle or close cluster.

Follicle, the fruit of a simple pistil opening by the inner or ventral suture.

Fruit, the matured product of the ovary with all its appendages; cf. Figs. 48, 74, 80, 83, 110, 115b, 116b.

Glabrous, bald, destitute of hair; cf. *smooth.*

Hypogynous, borne on the receptacle.

Imbricated, overlapping like shingles on a roof.

Incised, deeply and sharply cut as if slashed.

Indehiscent, not splitting open.

Inferior ovary, one which is not free; one adnate to the calyx.

Inflorescence, a flower-cluster, mode of flower arrangement.

Introrse, situated on the inside or directed inwards.

Involucre, a circle of bracts.

Involute, rolled in from each edge (Fig. 66).

Irregular, with the parts of different size or shape.

Lanceolate, lance-shaped, narrow and tapering gradually to a point (Fig. 62b).

Leaflet, one of the divisions of a compound leaf (Fig. 85).

Legume, a 1-celled pod opening by both ventral and dorsal sutures, like a pea pod. (Fig. 122).

Line, 1/12 of an inch.

Linear, 4 or 5 times as long as broad and with parallel or nearly parallel sides.

Membranous, thin, semi-transparent.

Moniliform, like a rosary (Figs. 71c, d).

Monoecious, with stamens and pistils in separate flowers on the same plant.

Nutlet, a small hard indehiscent 1-seeded fruit, usually derived from a compound pistil.

Oblique, developed more on one side than the other; not symmetrical.

Oblong, two or three times as long as broad and with parallel or tapering sides.

Orbicular, circular.

Ovary, the dilated or enlarged base of the pistil which contains the ovules.

Ovate, broad and tapering to a narrow apex; egg-shaped.

Ovoid, egg-shaped.

Ovule, the embryonic seed contained in the ovary.

Palmate, divided or lobed like the fingers of a hand (Figs. 70 and 79).

Panicle, a compound or branching raceme.

Parietal, placed on the side.

Pedicel, stalk of an individual flower or fruit.

Peduncle, stalk of a flower-cluster or cone.

Peltate, borne centrally beneath.

Perigynous, with corolla and stamens borne on the calyx.

Petal, a division or ''leaf'' of the corolla.

Petiole, the stalk of a leaf.

Pinnate, with the leaflets disposed along the two sides of a common axis (Figs. 85 and 120).

Pistil, the female organ of the flower.

Placenta, specialized tissue in the ovary which bears ovules.

Pollen, the fertilizing powder borne in the anthers.

Polygamous, with perfect and with either or both male and female flowers on the same or on different individuals.

Pome, a fleshy inferior fruit like an apple.

Pubescent, hairy with fine close hairs.

Punctate, dotted.

Raceme, having flowers with pedicels of about equal length disposed along a common axis and flowering from below upward.

Receptacle, the much abbreviated and modified stem which bears the various flower circles.

Regular, with the parts equal and of the same shape.

Reniform, kidney-shaped.

Revolute, rolled under or back from the edge.

Samara, an indehiscent pod with a long wing (Fig. 124b); double samara, two united pods, each with a long wing (Fig. 80).

Sepal, a division or "leaf" of the calyx.

Serrate, toothed like a saw with upwardly pointed teeth.

Serrulate, finely serrate (Fig. 85).

Sessile, without petiole, peduncle, or stalk; literally seated.

Sinuate, lobed with rounded recess (Fig. 49).

Smooth, not rough; cf. *glabrous.*

Spathe, differentiated bract-like leaf inclosing a flower-cluster.

Spike, a raceme in which the flowers are sessile.

Stamen, a male organ of the flower producing the pollen.

Staminodia, sterile stamens, usually scale-like.

Stigma, that portion of the style destitute of epidermis and fitted to receive and bring about the growth and development of the pollen-grains.

Stipitate, elevated on a slender stalk, as a stipitate gland.

Stipule, the appendages at the base of a petiole, one on each side.

Style, a slender often elongated organ connecting ovary and stigma.

Sub-, prefix meaning somewhat or approaching.

Subulate, awl-shaped.

Superior ovary, one which is free from or not adnate to the calyx.

Sympetalous, with united petals.

Synsepalous, with united sepals.

Terete, slenderly cylindric, circular in cross-section.

Thyrse, a contracted compact ovate panicle, one in which the middle branches are larger than those above and below as in the lilac and grape.

Tomentose, woolly.

Tomentum, wool.

Tree, a woody plant with distinct trunk and crown, commonly 20 feet high or more.

Truncate, cut off abruptly.

Two-lipped, with an upper and lower lobe to the corolla.

Umbel, a flat-topped flower-cluster with the pedicels of equal length and flowering from the outside towards the inside.

Umbilicate, with a central depression or umbilicus.

Umbo, with a central protuberance or point (the cone-scales in Figs. 12 and 13 have a prickly umbo).

Undulate, wavy.

Ventricose, puffed out or distended on one side.

Ventral, the side nearest the axis of growth; upper; superior; face. cf. *dorsal.*

Villous, with soft hairs.

OTHER BOOKS FROM CGR PUBLISHING AT CGRPUBLISHING.COM

1939 New York World's Fair: The World of Tomorrow in Photographs

San Francisco 1915 World's Fair: The Panama-Pacific International Expo.

1904 St. Louis World's Fair: The Louisiana Purchase Exposition in Photographs

Chicago 1933 World's Fair: A Century of Progress in Photographs

19th Century New York: A Dramatic Collection of Images

The American Railway: The Trains, Railroads, and People Who Ran the Rails

The Aeroplane Speaks: Illustrated Historical Guide to Airplanes

The World's Fair of 1893 Ultra Massive Photographic Adventure Vol. 1

The World's Fair of 1893 Ultra Massive Photographic Adventure Vol. 2

The World's Fair of 1893 Ultra Massive Photographic Adventure Vol. 3

Henry Ford: My Life and Work - Enlarged Special Edition

Magnum Skywolf #1

Ethel the Cyborg Ninja Book 1

The Complete Ford Model T Guide: Enlarged Illustrated Special Edition

How To Draw Digital by Mark Bussler

Best of Gustave Doré Volume 1: Illustrations from History's Most Versatile...

OTHER BOOKS FROM CGR PUBLISHING AT CGRPUBLISHING.COM

Ultra Massive Video Game Console Guide Volume 1

Ultra Massive Video Game Console Guide Volume 2

Ultra Massive Video Game Console Guide Volume 3

Ultra Massive Sega Genesis Guide

History of the Crusades Volumes 1 & 2: Gustave Doré Restored Special Editions

Chicago's White City Cookbook

The Clock Book: A Detailed Illustrated Collection of Classic Clocks

The Complete Book of Birds: Illustrated Enlarged Special Edition

1901 Buffalo World's Fair: The Pan-American Exposition in Photographs

Sinking of the Titanic: The Greatest Disaster at Sea

Gustave Doré's London: A Pilgrimage: Retro Restored Special Edition

Milton's Paradise Lost: Gustave Doré Retro Restored Edition

The Art of World War 1

The Kaiser's Memoirs: Illustrated Enlarged Special Edition

Captain William Kidd and the Pirates and Buccaneers Who Ravaged the Seas

The Complete Butterfly Book: Enlarged Illustrated Special Edition

Made in United States
North Haven, CT
05 November 2026

21763962R00135